MW00999514

LIC 2010

OTHER VOLUMES IN THE AUSA SERIES

Collins *Military Space Forces: The Next 50 Years*

Galvin *The Minute Men: The First Fight—Myths & Realities of the American Revolution*

Garden *The Technology Trap: Science and the Military*

Hilsman *American Guerrilla: My War Behind Japanese Lines*

Kinnard *President Eisenhower & Strategy Management: A Study in Defense Politics*

Kinnard *IKE 1890–1990: A Pictorial History*

Matthews & Brown *The Challenge of Military Leadership*

Matthews & Brown *The Parameters of Military Ethics*

Sajer *The Forgotten Soldier: The Classic WWII Autobiography*

Smith *Assignment: Pentagon—The Insider's Guide to the Potomac Puzzle Palace*

OTHER VOLUMES IN BRASSEY'S FUTURE WARFARE SERIES

Collins *Military Space Forces: The Next 50 Years*

Shaker & Wise *War Without Men: Robots on the Future Battlefield*

Simpkin *Race to the Swift*

Warden *The Air Campaign: Planning for Combat*

RELATED JOURNALS*

Armed Forces Journal International

Defense Analysis

Middle East Strategic Studies Quarterly

Survival

*Sample copies available upon request

AN AUSA BOOK

LIC 2010

Special Operations & Unconventional Warfare in the Next Century

Rod Paschall

Published with the
Institute of Land Warfare
Association of U.S. Army

Volume V
Perry M. Smith, *General Editor*

BRASSEY'S (US), Inc.

Washington · New York · London · Oxford
Beijing · Frankfurt · São Paulo · Sydney · Tokyo · Toronto

U.S.A. (Editorial)	Brassey's (US), Inc. 8000 Westpark Drive, Fourth Floor, McLean, Virginia 22102, U.S.A.
(Orders)	Attn: Order Dept., Macmillan Publishing Co., Front & Brown Streets, Riverside, N.J. 08075
U.K. (Editorial)	Brassey's (UK) Ltd. 24 Gray's Inn Road, London WC1X 8HR, England
(Orders)	Brassey's (UK) Ltd. Headington Hill Hall, Oxford OX3 OBW, England
PEOPLE'S REPUBLIC OF CHINA	Pergamon Press, Room 4037, Qianmen Hotel, Beijing, People's Republic of China
FEDERAL REPUBLIC OF GERMANY	Pergamon Press GmbH, Hammerweg 6, D-6242 Kronberg, Federal Republic of Germany
BRAZIL	Pergamon Editora Ltda, Rua Eça de Queiros, 346, CEP 04011, Paraiso, São Paulo, Brazil
AUSTRALIA	Pergamon-Brassey's Defence Publishers Ltd., P.O. Box 544, Potts Point, N.S.W. 2011, Australia
JAPAN	Pergamon Press, 5th Floor, Matsuoka Central Building, 1-7-1 Nishishinjuku, Shinjuku-ku, Tokyo 160, Japan
CANADA	Pergamon Press Canada, Suite No. 271, 253 College Street, Toronto, Ontario, Canada M5T 1R5

Brassey's (US), Inc., books are available at special discounts for bulk purchases for sales promotions, premiums, fund-raising, or educational use through the Special Sales Director, Macmillan Publishing Company, 866 Third Avenue, New York, NY 10022.

Library of Congress Cataloging-in-Publication Data

Paschall, Rod, 1935–
LIC 2010 : special operations & unconventional warfare in the next century / Rob Paschall.
p. cm. — (An AUSA book)
"Published with the Institute of Land Warfare Association of U.S. Army."
Includes bibliographical references.
ISBN 0-08-035982-5 :
1. Low-intensity conflicts (Military science) 2. Special operations (Military science) I. Title. II. Series: AUSA Institute of Land Warfare book.
U240.P38 1990
355.02′18—dc20 89-70807
CIP

Printed in the United States of America

This book is dedicated to my comrades in Delta Force, 1980–1982.

Brassey's Future Warfare Series

War will be with us for some time. Sadly we must recognize it to be one of mankind's most enduring endeavors. Our children must learn to defend themselves and so must their children. But if we prepare now, if we anticipate the nature of tomorrow's wars and begin to counter their new dangers, perhaps that future world—the world of the twenty-first century—will be a safer one than ours.

Brassey's has developed its Future Warfare Series for everyone concerned about tomorrow's world. National policymakers, military and civilian teachers, defense industry executives, informed citizens, and professional military personnel will all gain valuable insights through this series that will enhance their ability to meet the challenges of future warfare. Each book, written by an acknowledged expert, is intended to stimulate the reader's thinking, to raise new issues, and to initiate debate by peering into the future, boldly suggesting those factors that will shape warfare well beyond the twentieth century. Though every volume can and will stand on its own merit, our goal is to publish a series that encourages mankind to address the full spectrum of future warfare issues—now, before the future becomes the present and overwhelms the defenses of the past.

Contents

CHAPTER THREE

Special Operations and Low-Intensity Conflict Technology 41

CHAPTER FOUR

Special Operations in High-Intensity Conflict 67

CHAPTER FIVE

Special Operations in Mid-Intensity Conflict 81

Foreword

Understanding future warfare is the most important responsibility of those who must defend a nation from future enemies. Yet so many soldiers, scholars, journalists, and others concerned with defense policy tend to look to the future with their eyes fixed firmly on the rearview mirror. Fortunately, Rod Paschall understands the past, is a keen observer of recent trends, but most importantly has a seer's perspective on the long-range future.

Paschall has chosen to write about a portion of the warfare continuum that he knows very well. His experience as the commander of the famous Delta Force, an extraordinary group of American soldiers who are prepared to deal with international terrorism on a moment's notice, has served him well. His research and development assignments and his long tenure as the director of the Military History Institute at the Army War College have given him knowledge of emerging technology, as well as access to the historical record of special operations and unconventional warfare. His willingness to offer bold predictions makes this book much more than a tentative peek into an uncertain future. What Rod Paschall does in *LIC 2010* is simple but very important: he forces the reader to deal with the future of warfare, a future that will change significantly.

Why is low-intensity warfare so important for the future? The lessons learned from 1945 to 1990 are now quite clear. Nuclear deterrence has not only worked well, but mutual deterrence among the nuclear powers has also prevented potential aggressors among these nations from initiating conventional war against a major power. Why? No nuclear power wishes to engage in conventional war with another nuclear power for fear of nuclear escalation. As a result of mutual deterrence and other factors, low-intensity conflict has become the norm of international warfare and should continue to be the norm for many years to come.

As Paschall probes this area of warfare, he identifies some dramatic changes on the horizon. The reader should be ready for the mind-stretching experience of looking closely at rapidly evolving technology, three important special operations models, and changing doctrines. But most important of all, the reader should be prepared to deal with radical shifts in who will engage in insurgency operations and in which nations and regimes will be under insurgent attack.

My recommendation to the reader, then, is simple: if you are interested in the future course of conflict, inquire within.

MAJ. GEN. PERRY M. SMITH, USAF (RET.)
AUGUSTA, GEORGIA

Preface

This book is about the future of two distinctly different subjects, special operations and low-intensity conflict. If the reader is better able to separate the two after having read the book, one of its goals will have been achieved. If thoughts are stimulated by some of the forecasts, predictions, and calculated guesses, another purpose will have been fulfilled. Readers may differ on the wisdom of my effort, but there should be no dispute about the importance of the subjects, because they involve issues of freedom and repression, war and peace, life and treasure.

These pages are the product of much study and research, many years' experience in American special operations units, service in eleven Third World nations, and duties in the military research and development field. They have benefited from some help. Major General Perry Smith, Dr. Douglas Kinnard of the University of Oklahoma, Dr. Alex Roland of Duke University, and Lieutenant Colonel Jack Clark all plowed through early drafts. Later drafts were improved by comments from Benjamin F. Schemmer, editor of *Armed Forces Journal International;* Ron Hannon of Martin Simpson Company; Colonel Keith Nightingale; and William Rosenau of the John F. Kennedy School of Government, Harvard University. All are hereby held blameless for mistakes and errors. Acknowledgment must also be made to a little known but important group that I have labored with for the past five years, the Contemporary Military Reading List Panel of the U.S. Army War College. The panel, ably led by a fine soldier and scholar, Colonel Dave Hansen, chooses the best among hundreds of books each year. An author learns much from the panel's sometimes heated, but always enlightened and rational, deliberations.

CHAPTER ONE

Special Operations and Low-Intensity Conflict, 1990–2010

It is impossible to predict the future, yet almost everyone does it. A manufacturer buys materials to fabricate the goods he thinks will sell a month or year hence. A businessman orders the products he believes will be bought by the public during the next season. A banker forecasts future economic conditions and a client's behavior and makes or denies a loan. All would deny being clairvoyant, but each of them is staking fortunes on his perception of the future. They are analyzing trends. It is done every day.

The discussion that follows deals with a form of warfare, low-intensity conflict, and a type of military organization, specifically those units grouped under the general label of *special operations forces*. These two subjects are treated here not as they are, but as what they will be in the year 2010. The methodology is similar to that used by the manufacturer, businessman, and banker: an analysis of trends.

Barring the unlikely event of world peace, the occurrence of low-intensity conflict will unfortunately be far more prevalent in 2010 than it is today. However, the character of these small wars will be different. In the late 20th Century, these conflicts have often been marked by superpower intervention. In the future, the bipolar world will be greatly diminished, if not extinct, and external intervention will likely feature multiple actors, including commercial enterprises.

The incidence rate of terrorism may be in decline, but the lethality of the terrorist's weapons will likely grow, so that this component of low-intensity conflict will probably be more frightening in the next twenty years. The hopeful trend of growing international cooperation in the field of counterterrorism will probably continue, and the global community may see longtime adversaries joining together to fight terrorism, employing new techniques and technologies.

Insurgency, perhaps one of the oldest forms of warfare, should be much in evidence in 2010. But there will likely be significant changes in the next twenty years. In the 19th and 20th Centuries, the insurgent was largely content with using the most rudimentary weaponry and traditional techniques. By 2010, this will surely be altered in a dramatic way. The rapid and ongoing improvement in lightweight, shoulder-fired (or hand-held) anti-tank and anti-aircraft missiles, as well as other products of emerging technology, will make insurgents far more potent combatants in 2010 than they were in the late 20th Century. Guerrillas will be able to infiltrate into their enemy's territory with much more ease than their predecessors did. They will also have more secure means of communication, and they will be capable of being resupplied with relative speed and efficiency. Moreover, they will have low-bulk, lightweight materials and rations that will make insurgent groups more sustainable. These improvements, coupled with a political trend that began in the late 1970s, will probably transform the guerrilla fighter from a 20th Century tool of Marxist states into a major, 21st Century asset in the arsenals of Western powers. Western-supported and -sponsored insurgents will be used for the purposes of protecting and securing Western interests, and Western organizations that possess skill in supporting guerrilla forces will be held in readiness for purposes of deterrence.

The trend toward insurgents supported by democratic states will mean that counterinsurgency, once a near private preserve of the West and its clients, will be increasingly practiced by authoritarian regimes. It is unlikely that these nations will attempt to observe the legalities that some Western states have normally adhered to in their 20th Century anti-guerrilla campaigns. Technology will not favor the insurgent alone. In 2010, the counterinsurgent will probably employ area type weapons, munitions that "kill by the acre."

The first decade of the 21st Century should also see a considerable effort made in the field of peacekeeping. It will be one of the positive aspects of the coming century. Expanded peacekeeping operations will stem from a global condition relatively free of the past pervasive and inhibiting bipolar ideological confrontation.

Although there are likely to be many distinct differences between the late 20th Century and the early 21st Century, there should be some similarities. Special operations forces in 2010 will appear to be quite similar to the same types of forces of the late 20th Century. Yet, there will be changes. The most noticeable difference will be that there will simply be more of these units represented in the armed forces of the world. Still, they will likely constitute only a small percentage of any nation's defense establishment. It is also probable that these units will retain three basic organizational patterns, adaptations of the special operations formations of the United States, the Soviet Union, and the United Kingdom.

Additionally, it is unlikely that the fundamental missions of these forces—insurgency support, commando operations, reconnaissance, deception operations, psychological operations, military assistance, and counterterrorist activities—will be revised. However, in twenty years' time, special operations forces are quite likely to achieve unprecedented levels of performance based on several substantial technological advances and vastly improved personnel selection and training techniques. As in the past, these organizations will be prepared to conduct their missions in all levels of warfare: low-, mid-, and high-intensity conflict.

Definitions

Defining types of conflict, combatants, and units can be a mind-numbing endeavor, but unless terminology is generally understood, communication is all but impossible. Literature on both special operations and low-intensity conflict is often littered with emotion-burdened words and misleading descriptions. For example, the words *terrorist* and *patriot* have been used to describe the same individual, and elite infantry or even highly accomplished police organizations have been confused with special operations units. A foreign private citizen who assists a local insurgent group can count on being characterized as a "mercenary" by some reporter, as if the journalist were fully knowledgeable of the foreigner's innermost motivations. The term *low-intensity conflict* itself is understandably shunned or criticized by some Third World professional soldiers. After all, what seems to be a minor conflict to the industrialized world may involve national survival for the people of a developing country. What follows is perhaps a dry, but nonetheless dispassionate and objective, collection of definitions for words and phrases that are commonly used in discussions of both special operations and low-intensity conflict.

Low-intensity conflict is armed conflict for political purposes short of combat between regularly organized forces. That definition surely includes a terrorist act but excludes, for example, hostage-taking by a bank robber. It includes a counterinsurgency campaign in which a regular armed force is pitted against guerrillas or irregulars. It describes the activities of insurgents engaged in an armed attempt to overthrow a government. The definition also encompasses the efforts of a "peacekeeping force." Although commonly used, this latter term is on occasion one of the world's most flagrant misnomers, "peacekeeping forces" often being employed in some corner of the globe where there is no peace to keep. *Peacemaking forces conducting stability operations* is a more accurate definition and description of the forces and activity that are often found in conflict-resolution situations. However, when there is an actual condition of peace, an absence of any fighting, a force used to ensure continuation of that peace is accurately described as a "peacekeeping force."

Having defined what low-intensity conflict is, we may be prudent to define what it is not. It is not *mid-intensity conflict,* which is armed combat between regularly organized military forces. This definition describes the activity that most of the world's armed forces are organized, trained, and equipped to conduct. A mid-intensity conflict may include terrorist incidents, or even concurrent insurgent campaigns, but the main feature is the clash of two or more regularly organized military or naval forces. Traditionally, mid-intensity conflict has been the most certain way to either protect or defeat a government, a society, or a nation. However, mid-intensity conflict does not require the condition of two or more nations at war. Civil wars such as the American Civil War or the Spanish Civil War can be described as mid-intensity conflicts.

High-intensity conflict is armed combat involving the use of mass-destruction weapons. This type of conflict may include incidents or campaigns of low- or mid-intensity conflict. Since two atomic weapons were used in the last days of World War II, that event may be described as a high-intensity conflict.

These definitions differ somewhat from the officially blessed terminology of the U.S. Army, the organization that in 1975 decided to categorize warfare into three neat compartments.[1] However, they accurately reflect the intent of the U.S. Army definitions. This is important, because American soldiers have been increasingly successful in spreading the use of these conflict categories internationally. The above definitions are intended to be a bit less ambiguous and wordy than the ones in official American use.

Since the world's special operations forces are so diverse, it is better to describe them by what they do as opposed to attempting a precise definition of what they are. *Special operations forces* are those military or naval elements specifically organized, trained, and equipped to conduct or support insurgency, sabotage, psychological, deception, counterterrorist, foreign assistance, or commando-type operations. These units are normally capable of performing rescue, reconnaissance, and intelligence-gathering tasks. Special operations forces are not configured to conduct sustained combat against regular forces and are therefore not equipped with artillery, tanks, fighter aircraft, or combatant ships.

For example, the British Royal Marines are organized around "commandos," and one would therefore expect these units to be special operations forces. But the Royal Marine commandos have an artillery regiment permanently attached and are expected to conduct stand-up, continuous battle against an opponent's regular forces. The U.S. Marine Corps is much the same. It has no standing special operations force.[2] Similar reasoning can be applied to Soviet naval infantry and various parachute and light infantry formations of many nations—they may be commonly described as *elite units,* and may even occasionally conduct special operations–type missions, but they are not special operations units.

There is a tendency to confuse the relationship between special operations forces and low-intensity conflict. The tendency probably stems from the the prominence of U.S. Special Forces during the early days of the Vietnam War and the identification of the British Special Air Service with the counterguerrilla campaign in Malaya during the 1950s. The common error is to assume that special operations forces are the antidote to an insurgent force. That error leads to a false notion: the idea that special operations forces are primarily organized to conduct counterinsurgency campaigns. In reality, special operations forces are limited in what they can do in the field of counterinsurgency.

Critical tasks, normally essential for a successful counterinsurgency campaign, can be isolated and defined. They include population control measures, land reform, the establishment of a national-level intelligence collection and dissemination system, the training of infantry units to conduct effective small unit actions, and a host of political actions designed to undercut the insurgent's popular appeal. Although counterinsurgency task lists differ, a typical one includes about sixty generally recognized functions or tasks that should be performed if an insurgency is to be put down. Of those sixty or so tasks, U.S. special operations forces, the world's most diversified collection of such units, are only competent to accomplish eight or ten.[3] Special operations forces may be highly useful in a counterinsurgency campaign, but they are not the antidote to an insurgent movement.

Confusion is also often apparent in discussions where the terms *guerrilla, partisan, insurgent, terrorist,* and *mercenary* are used. *Guerrilla, partisan,* and *insurgent* are interchangeable. These three words refer to one who is part of an organization whose aim is to overthrow a government by armed force, largely through the use of indigenous resources. International conventions provide for the treatment of guerrillas, insurgents, and partisans. They must bear arms openly, wear an identifying symbol that is recognizable at a distance, and conform to the laws of war.[4] Compliance with these simple rules places the insurgent, guerrilla, or partisan in the category of a legally recognized combatant, one who is due prisoner-of-war status if captured.

Terrorists enjoy no legal protections. They normally conceal weapons, mingle with the civilian population for personal protection, and may take hostages to achieve their aims. Defying international conventions, they are usually treated as common criminals. Terrorist methods often involve armed and illegal, coercive propaganda.[5] The most typical terrorist goal is to achieve widespread recognition for a cause through outrageous actions that compel international attention.

One term, *mercenary,* is apt to be much in evidence during the early 21st Century, and it may be used as inappropriately then as it is now.

Commercial contractors currently maintain some weapons systems, perform housekeeping duties at military and naval installations, and conduct military training. They have even drafted military plans. The use of commercial firms in military affairs is growing, and their staffs are often composed of ex-military and -naval personnel. But are these commercial companies and their employees properly labeled as mercenaries?

The word *mercenary* is most often used in pejorative descriptions. The term usually has more to say about the writer or commentator's political orientation than it does about the person described. A true mercenary's sole motivation is financial reward, the acid test being whether he would switch sides for more money. In other words, the mercenary does not discriminate between political causes or nations to which he offers his services. His work simply goes to the highest bidder. As a practical matter, most people who are described as mercenaries are actually adventurers who discriminate between the political causes they support. A company that offers its military or naval services only to democratic nations and their allies could hardly be described as a mercenary organization. There are, of course, exceptions, but distinctions are becoming increasingly important, particularly in the realm of low-intensity conflict.

Special Operations Forces: 1990–2010

There are three general structures for the world's special operations forces: American, British, and Soviet. The armed forces of a number of nations pattern their own special operations units along the lines established by one of these three nations, and it is important to understand differences among the three models. Of the three, the American structure is by far the most complex. Two of these structures, the Soviet and American, are likely to undergo some degree of change in the twenty-year period 1990 to 2010. The least amount of change will probably be seen in the British model.

The British Special Operations Model

Two factors will produce a relative *status quo* for British special operations forces and the organizations that follow the British lead: the great success that the British have enjoyed with their special units during the 1980s and the paltry resources that London is likely to allocate to its military establishment during the next twenty years. Of the two factors, the potentially more damaging is the first. The victories that the 22nd Special Air Service (SAS) Regiment achieved under the brilliant leadership of Colonel Mike Rose during the 1980 counterterrorist operation at Princes Gate, the rescue operation in Gambia, and the regiment's excellent performance during the 1982 Falklands War all lead to the common British notion that revisions are unnecessary. A reputation for success tends to inhibit modernization, innovation, and change.

It can be expected that the organization, doctrine, and approximate strength of the British special operations forces in the year 2010 will remain about the same as they have been in the 1980s. All told, the British Army will have approximately one thousand active service members in special operations forces and another thousand or so in reserve. The largest contingent, about 750 strong, will be assigned to the 22nd Special Air Service Regiment of the British Army. The 21st and 23rd SAS Regiments, reserve or "territorial" army forces, will be slightly smaller. The Royal Marines will undoubtedly maintain the Special Boat Squadron, an element of less than 200 members.

The SAS structure for the 21st, 22nd, and 23rd Regiments will be organized around each regiment's four "saber" (or line) squadrons, units of about seventy men each. These squadrons will have four troops of twelve to sixteen men, organized into four-man teams. Team members will possess a variety of skills: parachute, medical, communications, language, and familiarity with a wide array of weapons.[6] Both regiments will be supported by a signals squadron and by supply and transportation units.

The Special Boat Squadron (SBS) of the Royal Marines will continue to specialize in coastal infiltration, reconnaissance, maritime raids, and maritime rescue operations. It is quite unlikely that the SBS will be able to keep up with its Soviet and American counterparts in the acquisition of coastal infiltration vehicles: mini-subs, helicopters, and small, high-speed surface boats. Unfortunately, this organization has been somewhat of a "stepchild." It has not enjoyed the prestige of the SAS and is not likely to be employed if the prospective mission is anywhere near a capability advertised by the SAS. Then, too, the SAS has developed facilities to train its teams in the fine art of coastal infiltration.

There is little reason to expect any major changes in the missions assigned to British specialized units. They will be tasked to perform counterterrorist, rescue, Third World military assistance, and low-level counterinsurgency tasks. The SAS will probably continue its occasional employment in the training of Third World protection services, particularly those elements providing physical security to chiefs of state. The SAS will also probably revive its former expertise in the field of insurgency support.

While past success stifles future improvement, there are facets of military operations that would only deteriorate if much change is made. In the British special operations model, a change in command and control arrangements would probably have negative effects. British special operations forces enjoy the world's best command, control, and coordination arrangements for specialized military formations. During periods of crisis or decision making that might possibly involve special units, the highest executive councils of British government include participation by a senior officer who is experienced in special operations. The officer is normally the director of Special Air Service, usually a former commander of the 22nd SAS Regiment. The capabilities and limitations of special operations forces are thereby made known to senior British political authorities at the very outset of the policymakers' deliberations. Although British intelligence services exercise no control over these special military organizations, close cooperation between special operations forces and the intelligence services is facilitated by high-level directives. The British system of command, control, and coordination for special operations units is one to emulate, not one to change.

There is, however, a discernible flaw in the British system. London has failed to provide its special units with a well-equipped air arm, particularly a long-range, air penetration means for use over hostile territory. Since most incidents requiring the use of special operations units during low-intensity conflict situations require a quick response, the SAS and the SBS often find themselves explaining their unique requirements to disbelieving air crews. Both units are likely to retain their status as "hitchhikers," dependent on whatever is available, not on what is actually needed.

The British model for special operations forces applies directly to similar organizations in the Australian, Jordanian, Omani, and New Zealand armed forces. To a lesser extent, the model is also applicable to several units in the South African military establishment, as well as a number of others belonging to the Third World states, former British colonies, or nations that have been influenced by British military traditions. However, the British pattern does not apply to East European Marxist countries or those Third World states that have been heavily influenced or supported by Moscow.

The Soviet Special Operations Model

An important difference between Soviet and British special operations forces is that the Soviet Union has a long-established and formal control link between its state intelligence and security apparatus, the KGB, and its military special operations forces, the Spetsnaz units. The political leadership of the Soviet Union is selected by a small group of career bureaucrats whose membership usually includes the KGB chief. Thus state policy is, in some measure, determined by the security and intelligence bureaucracy. Under such an arrangement, it is quite possible for Soviet special operations units to conduct operations without the knowledge or concurrence of some high-ranking Soviet military authorities. Such operations may be either internal or external to the geographic limits of the Soviet Union. The internal use of Spetsnaz forces is apt to be much in evidence during the 21st Century.

The late 20th Century has seen Marxism as a failed socio-economic system, one that must be imposed and sustained by force, and it is likely that Moscow will increasingly have to use force to maintain its power and influence. Spetsnaz units have a record of success in imposing "coups" on external states. Their "king-making" triumphs include Czechoslovakia in 1968 and Afghanistan in 1979.[7] There is no reason to believe they cannot be used to coerce or eliminate the leadership of dissident groups within the borders of the Soviet Union, to include reform-minded or recalcitrant military leaders. This type of employment sets the Spetsnaz apart from U.S. and British special operations forces.

Moscow has been particularly effective in its efforts to conceal information about its own military special operations forces and those of its allies. Western analysts have made estimates about Spetsnaz strengths, organization, and missions that have varied widely. Conservative projections put Spetsnaz strength at 12,000 active duty members, with possibly two to three times that number in a reserve status. The more sensational accounts have the Spetsnaz at 30,000—all highly skilled and poised for action.[8] The chief failing in the debate about numbers is that it overlooks what is possibly the most important facet in analysis of special operations forces: professional skill.

Both the ground and naval ranks of Soviet Spetsnaz forces are largely composed of two-year conscripts.[9] A large conscript force that can be rapidly expanded with reservists has been the standard for the Soviet military structure, one designed for a major ground war in Europe.[10] But it is unlikely that these Soviet special operations units could approach their British and American counterparts in levels of training and expertise. Short-term enlistment is not compatible with development of the skills required for special operations tasks. The conscript structure, useful for Soviet special operations forces in the 20th Century, is not likely to continue into the 21st Century.

It is probable that by the year 2010, Soviet Spetsnaz elements will have a strength of about fifteen to twenty thousand troops. These men may possess skills comparable to any of their Western counterparts. This force, above all others in the Soviet Armed Forces, will likely be a professional organization. And it is not likely that the future Spetsnaz organization will be dependent on short-term enlistees or the mobilization of reservists. The current assignment of some of the Spetsnaz units to field armies, theater commands, and fleets may be discontinued in favor of even more direct relationships with the KGB, the Ministry of Internal Affairs (MVD), and the Military Intelligence Directorate (GRU).

On the other hand, there is no reason to believe that the internal organization of Spetsnaz—eight- to ten-man teams within companies that contain fifteen teams each—will change. These companies will probably remain grouped in small brigades of about 700 to 1,000 men each. It is also likely that these units will obtain their own air arm before the 20th Century is out. The Spetsnaz air element will probably be a long-range helicopter force so that Spetsnaz is no longer dependent on "loaned" fixed-wing aircraft, used for parachute delivery.

The 21st Century Spetsnaz force will probably retain its commando heritage, but it will have to specialize in counterinsurgency operations as well as gain expertise in training allied units, both in Eastern Europe and in Third World Marxist states. Therefore, Spetsnaz officers will be plagued with the same dilemma that has worried their Western counterparts for many years: constant employment in low-intensity conflict tasks without relief from the responsibilities attached to mid- and high-intensity conflict missions.

It is likely that the four brigades of naval Spetsnaz will remain, and it is probable that their fine fleet of mini-subs and coastal infiltration boats will continue to be improved. The Soviets have a formidable development program for their naval Spetsnaz units, and the West can count on facing coastal infiltration craft of great speed, range, and carrying capacity.

The American Special Operations Model

Thus far, Soviet special operations forces have been limited to missions of reconnaissance, raids, and sabotage—the classic commando-type operations. In contrast, American special operations forces are expected to perform a multitude of tasks. These include military psychological operations, insurgency support, military assistance, raids, reconnaissance, counterterrorist operations, and even the conduct of military government functions, refugee affairs, and the handling of relations between military commanders and local populations in a battle zone.

To accomplish these varied functions, American special operations forces are spread across three services. The U.S. Army, Navy, and Air Force have an overall special operations strength of about 40,000 men and women in both active and reserve components.

Long-range penetration aircraft, both fixed wing and rotary wing, are provided by air force special operations elements. There are currently about forty such airframes in active service.[11] The Navy supplies thirty-seven SEAL (Sea Air Land) platoons, each composed of two seven-man squads. In order to infiltrate hostile coastal areas, the SEAL units have mini-subs, 14-man boats capable of 30 knots and having a 200 nautical-mile range, as well as "dry dock" devices for submerged launch of teams from submarines.

The largest contingent of U.S. special operations forces is assigned to the U.S. Army. A Ranger Regiment of three 550-man battalions specializes in raids, seizure of key facilities, and recovery missions.[12] There is also an active-duty psychological operations group composed of four battalions. The group is capable of broadcast, leaflet, and other uses of media in support of military operations. A number of other psychological operations groups are in a reserve status. In addition, almost all of the army's civil affairs capability is in the reserve structure. There is only one small, active-duty civil affairs battalion, a unit that, like its companion reserve units, mainly specializes in orchestrating local resources and manpower to support conventional forces.

To complement the air force capability for penetrating hostile air space in support of special operations, the army has a special operations aviation element with a number of rotary-wing aircraft. This relatively new unit has heavy, medium, and light helicopters, all specially configured for night penetration missions.

The most publicized element of America's special operations array is the army's Special Forces. Composed of four active and four reserve component groups, these units are not commando forces as they are often portrayed. They are specifically organized and trained to conduct insurgency support tasks and "foreign internal development" missions, a euphemism for counterinsurgency. Special Forces advise and teach indigenous forces. They are not organized or equipped to conduct combat operations themselves although, on occasion, they have done so. They have recently been given an additional mission, reconnaissance. Each Special Forces group is built around 54 "A" detachments of 12 men each. Detachment members possess expertise in medical, communications, demolitions, intelligence, and weapons subject areas. Each of the groups has a regional orientation. For example, the 7th Special Forces Group is dedicated to Latin America, and its membership is expected to be fluent in Spanish or Portuguese and must be familiar with Latin American customs.

A final element in the American special operations mix of forces is Delta Force. This unit is patterned on the model of the British 22nd Special Air Service Regiment. It is the only U.S. force that follows the European special operations pattern—it is a commando-type force.[13]

It is certain that the American special operations structure will undergo substantial change during the 1990s, change stimulated by the U.S. Congress. The American legislative body is acting with the general belief that these units are apt to be neglected by the American military services and with dissatisfaction over the performance of special operations forces in the 1980 Iran rescue attempt and during the 1983 operation in Grenada. Thus far, the U.S. Department of Defense and the American armed services have reluctantly implemented congressionally sponsored special operations initiatives, but there is little question that improvements have been made and little doubt that the U.S. Congress will continue to insist on further progress.[14]

Current plans envision the addition of another Special Forces group, plus a separate Special Forces battalion. Naval special operations forces are expected to double to about sixty SEAL platoons.[15] There may be a slight decrease in the large number of U.S. special operations components, as serious consideration is being given to the separation of Army psychological operations units so that their command and control level would be moved to higher echelons of the U.S. government. The most dramatic growth is scheduled to be centered on special operations air elements, within both the army and the air force.

During the late 1980s, over 250 U.S. aircraft were envisioned as being available to support American special operations forces by the mid-1990s. These aircraft were to include 55 new CV-22A tilt rotor airplanes, as well as over 90 updated versions of the venerable C-130, some of the latter being only partially earmarked for special operations tasks. Additionally, there was considerable support for converting two more former Polaris submarines to SEAL use, as well as making several Sturgeon class submarines available.[16]

It is doubtful that all this ambitious planning will see fruition. High costs, the probable decline in overall size of the U.S. armed forces, and the traditional reluctance of the uniformed leadership of the United States to support special operations enhancement at the expense of conventional forces will take a toll on plans for growth. However, these specialized forces are quite likely to be far more capable in the future than they have been in the past.[17]

It is also probable that the current confusing and unwieldy American special operations command and control arrangements will be revised by 2010. In contrast to the Soviet and British special operations models, which are largely centralized, the American pattern is badly fragmented. The root cause of the fragmentation is that American public law requires that combat operations be conducted by five regional, unified commands, scattered over the globe. This theater command system is troublesome enough in planning for the employment of the entire structure of the U.S. armed forces, but when these five commands each insist on their own share of the American special operations capability, a capability that represents only 1 percent of the U.S. armed forces, fragmentation rapidly depletes the available special operations forces. Then, too, missions for U.S. special operations forces may not always stem from strictly military requirements. For example, U.S. Special Forces have been employed by the Central Intelligence Agency in the past.[18]

Fragmentation also occurs along functional lines. The U.S. Special Operations Command has four component commands: the Joint Special Operations Command, the army's 1st Special Operations Command, the navy's Special Warfare Command, and the 23rd Air Force special operations unit. Each of these headquarters has some sort of operational function that requires a field communications element. When these communications requirements are added to those needed by the five unified commands, it is not surprising that there is not enough communications capability in any one place to conduct a single, substantial contingency operation.

The fragmentation of authority and responsibility does not stop with military and naval forces in the United States. There are nine different committees within the U.S. Congress that exert some measure of constitutional control over American special operations forces. Additionally, six different federal agencies within the executive branch of government have direct, statutory interests involved with the employment of special operations forces. It is unlikely that the American special operations command and control arrangements will survive a major test. It is likely that this system will see change, hopefully a change toward the centralization found in the Soviet and British models.

The American model for special operations forces has been followed by the armed forces of several Asian nations. The largest and most complete structures are those of Thailand, the Republic of Korea, and the Republic of China. Significant U.S. special operations influence exists in the Republic of the Philippines, Turkey, and in the armed forces of a number of Latin American countries.

Small Wars, Commandos, and the Like, 1990–2010

Trends in political events, economic shifts, demographic change, the constant march of technology, and evolution of relations between nations will shape the character of the next twenty years. It would indeed be unusual if those trends had no impact on the nature of insurgency, counterinsurgency, peacemaking, and terrorism. These forms of conflict are likely to experience some measure of change. It would also be peculiar if military organizations remained wholly static in the next two decades. They too will be altered.

In the late 19th Century, the incidence of *war,* defined as armed conflicts in which there were at least 1,000 deaths per year, stood at about 18 wars per decade. A century later, in the late 20th Century, the incidence rate was up to 27 per decade. Thankfully, the incidence rate per nation has decreased, there having occurred a threefold growth in the number of political entities. But the number of wars is increasing. There has been a dramatic rise in the incidence of low-intensity conflict. In the late 19th Century, 65 percent of all wars could be described as low-intensity conflicts. In the 1970s, low-intensity conflict represented 80 percent of all wars. A decade later, in the 1980s, low-intensity conflict accounted for 90 percent of all wars.[19] There is nothing readily apparent that is going to arrest this trend. It is a safe bet that the incidence of war will continue to grow, and it is likely that the great bulk of these conflicts will be low-intensity conflicts.

But there are trends within trends. The character of low-intensity conflict is changing. The incidence rate of terrorism may drop, but the lethality of weapons used by the terrorist may well grow. Insurgents will increasingly use more potent weapons, and their causes will increasingly coincide with Western interests. The counterinsurgent will likely be Marxist or represent some other form of totalitarian or authoritarian rule. Finally, there is apt to be considerable military activity in the realm of stability operations, peacekeeping and peacemaking.

In the late 19th Century, there were practically no military organizations that could be described as special operations forces. There are now hundreds of such units, and it is the rare nation that does not sponsor and support these formations. They are not solely dedicated to employment in low-intensity conflicts, but they are often useful in that level of warfare. The world's special operations forces are, as noted, being developed along three general paths, azimuths that are being set by the British, American, and Soviet armed forces.

The British special operations structure is likely to reflect little change over the next two decades, but there should be a shift toward a better capability in the field of insurgency support. Soviet special operations forces can be expected to be increasingly dedicated to counterinsurgency tasks, and it is probable that these elements will themselves follow the American lead in acquiring an air arm. As for the latter forces, they will hopefully improve their command and control arrangements, achieving a larger measure of centralization. The U.S. special operations forces are also likely to be heavily involved with sponsoring, supporting, and training the insurgents of the early 21st Century.

The nature of conflict, whether that conflict is of low, mid, or high intensity, shapes the character of forces that conduct military operations. But conflict, too, is shaped. It is shaped by politics. Politics, in turn, is influenced by a number of factors, factors that include demography, economics, ideology, and the desires and expectations of people. Although it is impossible to determine with any degree of precision what the politics of 2010 will be, it is possible to define trends that are working to effect political change.

Notes

[1] Interview with Stanley Hyrowski, Kennedy Center for Special Warfare, Ft. Bragg, North Carolina, December 1982. Hyrowski was the author of the U.S. Army's doctrine, *Field Manual 100-20: Low Intensity Conflict.* The manual contained the first official U.S. definition and use of the term *low-intensity conflict.*

[2] Lieutenant Colonel D.R. Blankenship, "Marine Expeditionary Unit (Special Operations Capable)," *Amphibious Warfare Review* (Summer 1988): 43–49.

[3] Author's personal experience as a 1983 low-intensity conflict study team leader. The study determined, among other things, the utility of U.S. special operations forces in the field of counterinsurgency.

[4] Geneva Conventions, August 12, 1949, Geneva, Switzerland.

[5] Francis M. Watson, *Political Terrorism: The Threat and the Response* (New York: Robert B. Luce Co., 1976), p. 15.

[6] Tony Geraghty, *Inside the SAS* (New York: Ballantine Books, 1982), pp. 248–250. Ross S. Kelly, *Special Operations and National Purpose* (Lexington, MA: Lexington Books, 1989), p. 37.

[7] Albert Seaton and Joan Seaton, *The Soviet Army: 1918 to the Present* (London: The Bodley Head, 1986), p. 228.

[8] Viktor Suvorov (pseudonym), *Inside Soviet Military Intelligence* (New York: Macmillan Publishing Co., 1984), pp. 7–11.

[9] Viktor Suvorov, *Spetsnaz: The Inside Story of Soviet Special Operations Forces* (New York: W.W. Norton Co., 1987), pp. 67–84. David Isby, "Soviet Special Operations Forces in Afghanistan, 1979–1985," *Report of Proceedings: Light Infantry Conference, 1985* (Seattle: The Boeing Corporation, 1985), pp. 182–186.

[10] Robert S. Boyd, "Spetsnaz: Soviet Innovation in Special Forces," *Air University Review* (November–December 1986): 63–69.

[11] Major Bradley J. Baker, "Air Force Special Operations: How Did We Decide What Was Enough?" *Airlift* (Spring 1988): 14–17.

[12] Lieutenant Shaun M. Darragh, "Rangers: The Long Road to Recognition," *Special Warfare* (April 1988): 19–27.

[13] John M. Collins, *Green Berets, Seals, and Spetsnaz: U.S. and Soviet Special Military Operations* (McLean, VA: Pergamon-Brassey's, 1987), pp. 12, 23, 26–27, 31. For the development of the Delta Force, see Charlie A. Beckwith and Donald Knox, *Delta Force* (New York: Harcourt Brace Jovanovich, 1983).

[14] Lynn Rylander, "Congress Takes Action to Modernize Forces for Special Operations," *Amphibious Warfare Review* (Summer 1988): 60–65. Letter from U.S. Senators Sam Nunn, Edward M. Kennedy, John W. Warner, and William S. Cohen to Lieutenant General Brent Scowcroft (USAF, ret.), assistant to the president for national security affairs, dated January 25, 1989, as quoted in *Armed Forces Journal International* (March 1989): 66–67. Lieutenant Colonel William Cowan (USMC, ret.), "Strike Back? Just as Soon as the Paperwork Is Ready," *Washington Post,* April 23, 1989, p. C1.

[15] Norman Polmar, "SOF—The Navy's Perspective," *Proceedings* (August 1987): 136–138.

[16] Frank C. Carlucci, *Annual Report to the Congress* (Washington, D.C.: Department of Defense, 1988), pp. 225–230.

[17] General James Lindsay, commander in chief, U.S. Special Operations Command. Remarks to the U.S. Senate Armed Services Committee, April 20, 1989, as quoted in Rick Maze, "Dropping V-22 Could Limit Special Operations," *Army Times* (May 8, 1989): 26.

[18] Jeffery J. Clarke, *United States Army in Vietnam, Advice and Support: The Final Years, 1965–1973* (Washington, D.C.: U.S. Army Center of Military History, 1988), p. 69.

[19] Ruth Leger Sivard, *World Military and Social Expenditures, 1987–1988,* 12th ed. (New York: World Priorities, 1987), pp. 28–31.

1990s, U.S. bases may no longer exist in Spain or Greece, and it is likely that greater restrictions will be placed on U.S. use of Lajes Air Force Base in the Azores. The majority of U.S. registered voters believe America's defense of other nations is a threat to its own security, a reversal of the 1960s attitude.[9] A poll of U.S. opinion makers has identified the threat of nuclear war as the most serious projected concern in the year 2000, while a broad sample of U.S. opinion indicates a 75 percent preference for a cooperative problem-solving relationship with the Soviet Union as opposed to U.S. military superiority, a preference that merited only a 39 percent approval rating. Additionally, polling data throughout the mid- and late-1980s indicated America's public desire for smaller defense budgets.[10] All of these factors work to weaken military ties between the United States and its West European allies.

The effect of these attitudinal changes may be compounded by demographic trends within the United States. The rapid decline in the size of the average American family reached new lows during the 1970s, producing smaller numbers of military age youths for the 1990s. The rise in the U.S. median age is projected to continue into the 21st Century, and may well contribute to a more conservative and isolationist American public attitude. Additionally, European-oriented states such as New York, Pennsylvania, Michigan, and Massachusetts are losing population and representation in the Congress to those states that are Pacific and Latin America–oriented—Texas, California, Florida, and Arizona. These latter states may control over 30 percent of the U.S. House of Representatives in the year 2010.[11] None of these demographic trends, or the growing numbers of Hispanic and Asian immigrants to America, make for better bonds between the United States and its European allies.

The certainty of diminishing global oil supplies, however, is apt to pose the greatest single obstacle to U.S.-European unity in the late 1990s and early 21st Century. Both of these heavily industrialized societies are extremely dependent on the availability of cheap oil, and the United States may end its own oil production by the year 2020. With little growth in nuclear power, and increasing environmental concerns over the use of coal, U.S. oil consumption is rapidly rising once again. Falling from about eight to five million barrels per day between 1977 and 1985, the American appetite for foreign oil grew during the late 1980s, reaching the earlier 1977 levels at the end of the decade. By the mid-1990s, American imports of oil will probably be at twelve to thirteen million barrels per day. There are reports of growing world oil reserve figures, but the larger numbers come from the Organization of Petroleum-Exporting Countries (OPEC) states, which have an interest in higher sales quotes—quotes based in part on reported reserves. There is, however, no doubt as to the increasing dependence of both Europe and the United States on Middle East oil.[12]

The events of the Arab-Israeli War of 1973 proved that Western Europe will not jeopardize its access to Middle Eastern oil producers for the sake of supporting U.S. security policies, a position that is not likely to change. Barring technological breakthroughs that would reduce the enormous U.S. need for foreign oil, America may be in a serious state of dependency in the 21st Century. Moreover, U.S. and European competition for increasingly scarce and expensive oil may well create a confrontational atmosphere within the West. A more energy-efficient Western Europe is likely to consolidate its oil-buying power in order to bargain more effectively against U.S. needs.

Another factor that indirectly produces a widening gap between North America and Western Europe has to do with the unceasing proclivity of the United States to produce and use the most sophisticated weaponry for the European defense role. In contrast, European nations habitually appear to be satisfied with a slower pace of modernization. The American policy stems in part from the understandable need to offset the superiority of Warsaw Pact numbers with a U.S. qualitative advantage and in part from the ready availability of a technologically advanced American arms industry. A contentious facet of this U.S. policy springs from American pressures to sell sophisticated weapons to European allies in order to cut unit prices to the U.S. taxpayer. However, there are distinct limits to this preferred American policy.

In the 1950s, the United States spent about 11 percent of its gross national product on defense, a figure that was reduced to about 8 percent in the 1960s and less than 6 percent in the late 1980s. The growth of the American economy has offset these reduced defense burdens during much of the late 20th Century, and thus one would expect substantial increases in relative combat power due to the rise in America's GNP. But that GNP growth has not been substantial enough to keep up with the rising costs of U.S. weaponry. While real growth in the U.S. economy has been at a rate of about 2 or 3 percent during good years, average prices of conventional American weapons such as tanks and combat aircraft have climbed steadily at a rate of about 5 percent each year.

Moreover, the overall U.S. technological and financial lead is slipping. As recently as 1974, Americans created as much as 70 percent of the world's advanced technology. Ten years later, that figure was down to 50 percent, and in 1986 the United States ran its first trade deficit in advanced technology products.[13] This decline has affected the U.S. defense industry, where America is rapidly losing its lead in semiconductors, computers, numerically controlled machine tools, and precision optics.[14] Considering the steady improvement in Soviet arms, the rising costs of American weaponry, and the steady erosion of the U.S. technological lead, Washington's preference for qualitative weapons superiority may be a policy of the past by the mid-1990s. By 2010, it is projected that Moscow will attain a superiority in a combined qualitative and quantitative index for nine of the fourteen categories of major weapons systems. The United States is expected to lead in only three of the systems—bombers, land attack cruise missiles, and naval surface combatants.[15] Additionally, there is an ongoing relative reduction in America's financial strength. In 1970, seven U.S. banks were among the world's top twenty-five in terms of assets. By 1987, only one American bank ranked in the top twenty-five. In a relative sense, the United States is rapidly losing its technological and financial leadership.

A new West is beginning to emerge. By the late 1990s, the post–World War II arrangements that existed among the Western democracies for a half century will have been transformed into a bipolar relationship. One pole will be in Washington, with the other focused at the center of a complex web of West European economic, political, and security agreements. The European community is becoming an increasingly cohesive confederation of democratic states. Halving the West, however, does not necessarily produce weakness. Both Western Europe and North America are growing in strength at a much faster rate than most other regions of the world. It is just that some of America's great lead in trade and technological leadership has been cut marginally, partially due to European gains. The emerging European confederation's independence from Washington is thus the natural consequence of economic and technological success. While there will be a considerable degree of competition in the New West, particularly in the area of trade, common interests will ensure occasional cooperation. This dual alliance will have much that will bond it; non-Western nuclear proliferation, environmental concerns, human rights issues, and global economic stability are just a few of the issues that will find Western nations on common ground. And the New West is also likely to present a unified response to an old concern—any serious threats posed by Moscow.

The East

While the West is being transformed into two focal points of power—North America and a West European confederation—the East will likely continue under the influence of three great nations: China, Japan, and the Soviet Union. In some respects, the New West will enjoy an advantage over the East. In periods of crisis, the two western centers may form a formidable, albeit temporary dual alliance. There is little likelihood of such cohesion in the East. The three eastern powers will remain quite dissimilar, possess few common interests, and probably retain considerable mutual distrust. However, the appearance of *status quo* in the East is only superficial, because two of these three great eastern countries are in the throes of rapid change.

Substantial change, at least in the somewhat narrow field of security policy, cannot be expected in the late 20th Century and early 21st Century for Japan. This industrial and technological giant has already overtaken the Soviet Union in terms of gross national product, leaving the U.S.S.R. as the world's third largest economic power. Subtracting the imported wealth of Japan, the money earned by its foreign trade, the resulting gross domestic product of Japan is actually less than that of the Soviet Union. Few nations are more dependent on foreign trade to underpin a standard of living, and therein lies the reason that Japan is unlikely to revise its carefully balanced security posture, a mixture of pacifism, a subordinate, low-key role in its security arrangement with the United States, and a militant insistence on keeping the whole focus of that bilateral treaty on the defense of its home islands. Japan will go to almost any lengths to ensure that its security policy does not interfere with its trade policy. It has been immensely successful in this endeavor in the past, and there is no reason to expect that it cannot continue to succeed in the future. Geography, the United States, and Japanese will power are on Tokyo's side.

Japan's neighbor, China, is expected by some expert observers to exercise far more power and influence in the early 21st Century than it has thus far done in the 20th Century. A simple, straight-line projection of Beijing's fast-rising gross national product will put it at $3.7 trillion in 2010. A similar calculation produces only $2.9 trillion for the Soviet Union, and that means Moscow could well lose out once again, falling to the world's fourth economic power behind the United States, Japan, and China.[16] Unlike Japan's growing economy, China's rising economic strength is likely to lead to a significant change in the security arena of the early 21st Century.

Should the United States and the Soviet Union agree to reduce their strategic nuclear arsenals by 50 percent, a likely event in the 1990s, China will then possess a strategic nuclear stockpile 30 percent the size of that controlled by either Washington or Moscow. By simple reason of magnitude alone, China would then become a more important actor on the world stage. The lineup of major nuclear powers would then be the United States and the Soviet Union, followed by China, Britain and France, then India. The Chinese, of course, are contiguous to two of these nuclear powers and have long-standing territorial disputes with both.

Not only will Beijing be a nuclear power to reckon with, it will have a much improved capability to project military strength in pursuit of its interests. Twentieth Century strategic thinking about China has invariably involved a number of oft-repeated considerations that limited the supposed importance of the Chinese. With a population of over one billion, or one-fifth of the world's people, China has been viewed as a nation whose fundamental social burdens precluded it from undertaking most foreign military endeavors. But its population control measures of the 1980s, while not altogether successful, have resulted in over a hundred million fewer births, and its agricultural and industrial progress is producing a rather bright future for the ancient society. China has created a thriving arms industry, one that has been used to mechanize its own armed forces and contribute to Beijing's export earnings and foreign security interests. Past assumptions about China's slight military technological base are rapidly disappearing with each new shipment of Chinese-made missiles and combat aircraft to Third World states. Although the turbulence of the spring and summer of 1989 have raised the level of uncertainty, the China of the 21st Century should be considered as a first-tier military power.

The third power of the East, the Soviet Union, is being influenced by a wide range of internal changes, but the direction of transformation and probable status twenty years hence is difficult to discern. There are, however, a few known factors that will contribute to the shape of Moscow's future: population, ethnic composition, the record of Marxism, and the Soviets' relations with their allies and clients.

Compared to an expected U.S. population of 280 million in the year 2010, the Soviet Union's population is projected to reach 336 million in that year. Both nations will experience ethnic changes, but the ethnic composition of the U.S.S.R. in 2010 poses more problems for Moscow than changes in the United States. The Great Russians, who largely reside in the biggest of the fifteen republics, the Russian Federated Socialist Republic, are surrounded and steadily being outnumbered by ethnic minorities, particularly by the Moslem peoples of the U.S.S.R. By the year 2020, half of all births in the Soviet Union will be to Moslem families, families which are, on average, three times larger than Russian families. In fact, population trends clearly point to the high probability that the Russians themselves are destined to become a minority, comprising only 48 percent of the Soviet Union's population at the end of the 20th Century. In that year, one-third of all recruits in the Soviet Armed Forces will be from the Islamic republics, and there will be 78 million people on Moscow's troubled southern tier in Armenia, Azerbaijan, Turkmenistan, Kazakhstan, Uzbekistan, Kirgizia, and Tadzhikistan. Ten years later, that figure will jump to 92 million, or 27 percent of the U.S.S.R.'s population.

This ethnic revolution would be of slight concern to Moscow if Soviet society had the American "melting pot" system of assimilation. The U.S.S.R. has simply not developed in that way, however, and it now has 90 to 120 significant language groups, 22 of them having over one million speakers each. Already, almost 25 percent of Soviet soldiers lack fluency in the Russian language, and their leaders are beginning to complain that recruits do not understand the oath they take on entering service. Some officers are requesting interpreters. In the broader society, the ethnic Russians have maintained a firm grip on leadership, holding 68 percent of the top-level political positions and similar percentages in the higher-paying professions, crafts, and trades. On the whole, most non-Russians are relegated to agricultural and common labor pursuits. The Great Russians are sitting on a demographic powder keg that each year becomes a bit more unstable.

There is, however, one ray of hope for Moscow in the otherwise bleak realm of Soviet demography. All N.A.T.O. countries will experience declines in military age youth by 2000, while all Warsaw Pact nations except East Germany will see an increase in the number of 18- to 20-year-olds. For example, the Federal Republic of Germany considered extensions of service for its conscripts to compensate for an expected 40 percent drop in available West German youth. During the same period, Poland will have a 30 percent increase in its draft eligibles. But, things being what they are with the Soviet Union's western satellites, Moscow, were it given the choice, might choose to exchange the demographic projections of Poland for those of West Germany.

There are many reasons for the disenchantment of the peoples of the Soviet western republics and the citizens of the Warsaw Pact nations, but a chief cause for unrest is that Marxism-Leninism has not delivered on its promises. Despite the wholesale uprooting of entire societies, massive labor projects, great sacrifice, and vast expenditures of resources, energy, and talent, European communist nations and former European nations that have been incorporated into the Soviet Union have experienced lackluster economic performance, marginal social gains, and non-stop political repression. Progress—what little has occurred in such nations as Hungary—is widely attributed to deviation from the principles of Marxism and not to adherence to those principles. The great expectations that communism engendered—economic improvement, elimination of class rivalry, and the promised disappearance of the state—have long since been crushed under the numbing weight of massive party and government bureaucracies, organizations that stifle innovation and initiative. Marxism has not delivered; the great socialist revolution has been an abject failure.

Although Moscow's leaders will have to keep a wary eye toward the Moslem south, their greatest concern in the 1990s is apt to be west of the Russian Soviet Federated Socialist Republic. Internally, Estonia, Latvia, and Lithuania are straining an already acrimonious relationship with Moscow, pressing for any possible opportunity of independence. In 1988, the Byelorussian representatives to the Supreme Soviet openly opposed the Politburo's plan for a new tax structure, a rare event of party indiscipline. During the late 1980s, the Kremlin was challenged on a wide variety of issues from the Ukraine, Georgia, the Crimea, and other sectors of the U.S.S.R. Externally, Polish unrest threatens the geostrategic stability of the Warsaw Pact. If the Russians do have to contend with internal revolt or serious external resistance in the 1990s and early 21st Century, the flame may spread to the south, but the spark will likely come from socialist republics and nations west of Moscow.

Perhaps the current "restructuring" campaign will defuse the potentially explosive conditions within the Warsaw Pact nations and restive Soviet republics. The end result is impossible to predict, but there is little doubt that, once again, expectations have been raised in the communist world. Progress, or the lack thereof, will be evident by the mid-1990s, and Moscow will either enjoy revitalization due to restructuring or suffer the consequences. Either way, global security conditions are certain to be altered.

While social, political, and economic progress have been hard to come by in Marxist states, improvement of their military and naval forces has been substantial. The large Western qualitative lead in tanks, aircraft, naval combatants, missiles, anti-aircraft systems, and communications equipment steadily eroded during the 1970s and 1980s as the Soviet Union and its allies fielded generation after generation of modernized weaponry. In the face of poor economic growth, Moscow dug deeper, raising the defense share from 15 percent of the gross national product in 1980 to 16 percent in the late 1980s.[17] Much of this expensive modernization and refurbishment is being accomplished at a time when the West is decreasing its outlays for conventional forces.

Although there are likely to be East-West negotiated reductions in conventional arms during the early 1990s, it is doubtful that Marxist states will surrender much of their advantage. There is a high probability of a continuing superiority in Warsaw Pact numbers and conventional munitions stockpiles over the allied nations of the West. Communist nations are simply more dedicated to military strength than are democratic nations. Thus in the 1990s it is likely that there will be virtually no remaining Western military advantage over the Marxist coalition other than the professional competence of N.A.T.O.'s soldiers, airmen, and sailors. The likelihood is that the military technology gap will be narrowed, if not closed.

There is one other area of Marxist strength. Marxist nations have no peers in the realm of internal security organizations. The KGB, the MVD, and the foreign security services that these two organizations have sponsored are well trained and massive but not necessarily effective. Marxist security services are so pervasive and active that they create opposition. Communist states that have effected substantial liberalization—China, Yugoslavia, East Germany, Rumania, and Hungary—have usually done so from the top down. The choice was largely made by Marxist leaders and not by the armed force of an internal opposition. Marxist security services have a nearly unblemished record of keeping a ruling elite in power—if that elite chooses to stay in power. That characteristic makes such organizations attractive in some quarters of the Third World.

The Third World

Although esoteric classification schemes have been developed to differentiate between various categories of the globe's poorer nations, most people readily recognize the term *Third World* as referring to the 113 preindustrial nations located in the southern hemisphere and lower tier of the northern hemisphere. These states are unfortunately becoming increasingly authoritarian, armed, populated, poor, and engaged in conflict. The term *developing world* is no longer appropriate because about one-third of these countries are experiencing economic regression, their people growing hungrier and more destitute with each passing day. During the late 1990s and early 21st Century, the Third World will undoubtedly see the vast majority of all low-intensity conflicts, and this region will continue to be an area where the special operations forces of many nations are employed.

In 1960, 26 percent of all Third World states had key government positions held by military officers, extrajudicial authority exercised by security forces, and a lack of civilian control over their armed forces. Despite several successful liberal reform efforts in Latin America during the 1970s and 1980s, the percentage of authoritarian nations in the Third World doubled, reaching 52 percent in 1987. Authoritarian government is particularly prevalent south of Africa's Sahara Desert, where only 7 of 45 countries have multi-party political systems. These autocratic regimes have twice the military expenditures of other Third World states, and their ruling cliques have extended power tenures that average 19 years.[18] Third World martial governments often engender armed insurgency, the most bloody form of low-intensity conflict. If the long-term growth trend for authoritarian government holds, there will be more insurgencies in the 21st Century than there were in the 20th, and there will be a greater call on the advisory services of the Soviet KGB and MVD by either right- or left-wing Third World autocracies besieged by guerrillas.

As the Iran-Iraq War proved in the 1980s, Third World states are capable of conducting devastating mid-intensity conflicts, and that capacity is growing. In 1960, at the outset of independence for many of these countries, few Third World states possessed the ability to attack a neighbor, consolidate a victory, or even defend themselves. By the 1980s, about one-third of these nations had attained such capabilities.[19] In the late 1970s, arms imports by the Third World doubled, reaching $35 billion.[20] During the 1980s, the Third World increased its arms imports, buying $180 billion worth of weaponry in 1987. Some of these states have created their own defense industries. Arms sales from other Third World nations constituted 6 percent of Third World military procurement in 1980, and that figure rose to 11 percent by 1986. Nineteen Third World countries now have ballistic missiles with typical payloads of 1,100 pounds and ranges from 75 to 1,600 miles, and it is expected that fifteen of them will be manufacturing such weapons in the year 2000.[21] The Third World is arming well and quickly.

Another long-term and unmistakable trend is the gradual domination of Third World arms transfers by the Soviet Union. The U.S.S.R. accounted for 35 percent of the world's weapons transfers in the late 1980s, a rising figure, but the real growth has been in its supply of arms to poor nations. In the 1930s, Moscow handled about 6 percent of this trade. By the 1960s, it was responsible for 30 percent, and by the mid-1980s the Soviets were the delivering party for over 50 percent of the transactions. An obvious objective of this Soviet activity is gaining influence in a supposedly resource-rich south. But, it is quite unclear if there is a direct correlation between the policies of the arms supplier and those of the receiving state due to arms transactions, subsequent technological advice, and assistance. Third World nations now have wide choices among arms suppliers, and they have entrenched political organizations. They are therefore far less fragile and susceptible to external influence than they were in the early days of independence. However, persistent failure has never been an impediment to Moscow, and a thriving arms trade between the Soviet Union and the world's poorer nations can be expected well into the 21st Century.

Poverty and hunger do not necessarily breed war. Most of the world's poor bear their economic misfortune stoically. But, poverty and hunger are used by insurgent leaders as motivations to take up arms. There will be many opportunities for those calls in the 21st Century. The Third World has 75 percent of the world's population, but only about 25 percent of the globe's combined gross national product, and the gap between the few "haves" and the many "have nots" is widening. By 2010, the world's population will have grown from 4.9 billion in the mid-1980s to 7 billion, 80 percent of whom will be citizens of the Third World. Although it is estimated that current technology and agricultural techniques could comfortably feed 6 billion people, 10 percent of mankind is at or near starvation. A part of the reason lies in agricultural disincentives practiced by Third World governments that subsidize the city dweller by keeping food prices low. Conversely, Western nations subsidize the farmer. Latin America, a net grain exporter in the 1970s, became a net grain importer in the 1980s, taking 10 percent of North American grain exports. Africa, once a continent that easily fed itself, is now importing 40 percent of its foodstuffs and its farmers are increasingly deserting their fields.[22] By the year 2000, 42 percent of Africans and over 60 percent of all Latin Americans will be urban slum dwellers. In Asia, the Republic of the Philippines is expected to double its population to 120 million by 2010, and 10 percent of all Filipinos are destined to face an uncertain existence in the hovels of Manila. In addition, Third World populations are rapidly attaining a new characteristic in that their median ages are rising. In the late 1980s, 58 percent of the world's 55 years-plus age group was in the Third World. By 2010, 67 percent of this age group will be there, citizens that need medical treatment for chronic illnesses, the most expensive categories of maladies.[23] A century ago, rapid population growth, lengthened life spans, and urbanization were descriptive of the United States, but the Third World lacks the other 19th Century American characteristic, fast-paced economic development.

Unfortunately, there is little hope the Third World can reclaim the title of "developing world." Development requires investment, educated talent, and a market environment conducive to economic growth. It is likely that the Third World will remain short of all three. The current trillion-dollar plus Third World debt is regarded as lost by most investment bankers. In the 1990s, Western banks conducting international operations will require 4 percent equity backing. In order to compete, many banks are selling their Third World loans for fifty cents on the dollar or less. While this may provide temporary relief for some of the borrowing countries, it is unlikely that any of the major Western financial institutions will recreate the conditions they put themselves in during the 1970s and 1980s. Except for some short-term loans, the Third World is likely to be drastically capital poor during the early part of the 21st Century. In forty of these countries, education is being reduced, not enhanced, governments there cutting per capita funding for education by as much as 25 percent.[24] Additionally, there is no indication that Third World nations will be able to stem the tide of their "brain drain," the continuing exodus of their educated class to the West. Lastly, it is now apparent that development within the Third World is unlikely to be externally stimulated, and that what economic development is achieved is often done in an illegal fashion that pits Third World entrepreneurs against their own governments. Despite almost forty years of effort and billions of dollars expended, the U.S. Agency for International Development has finally concluded that only a handful of Third World countries have achieved economic independence.[25] Another study has concluded that both left- and right-wing bureaucracies in the Third World are inherently hostile to beginning indigenous businesses and industries. Progress in these enterprises, when it is achieved, is often accomplished only by bribes, circumvention of laws and, in some instances, outright violence.[26] Rapid, peaceful economic development does not appear to have a future in much of the Third World.

A final discouraging trend in the Third World is the rising propensity to use force in an effort to achieve political goals. Since World War II, almost all wars have been in the Third World, and most have been low-intensity conflicts, armed violence short of that between regularly organized armies. Not only has the incidence of war been growing, the length of wars has also been on the increase. By 1985, the average length of a war was five years. However, war is now a doubtful proposition as an instrument to effect political change. Up to the 20th Century, the initiator of war was likely to win. Now, there is only a 21 percent chance for the initiator to win, a 34 percent chance of losing, and 45 percent probability of reaching only a draw.[27] Violence, however popular it is becoming in the Third World, offers small hope of bringing about positive results.

A World of Diffusion

Some trends that will probably have an impact on security issues in the early 21st Century are not regional but universal in nature. Three in particular appear to have relevance in the world of special operations and low-intensity conflict: the pervasive desire to emulate Western lifestyles, the increasing use of the English language, and—a more ominous one—the proliferation of nations capable of producing and using nuclear and chemical weapons.

For good or ill, there is a near-universal quest for the highly materialistic Western mode of living. With few exceptions, single-family homes, personal automobiles, TV sets, videocassette recorders, and home computers are prized, even in remote regions of the world. Western magazines, movies, and TV productions penetrate almost every society, stimulating desires and creating an ever-expanding market for Western innovations. Although this factor may constitute an obstacle between a Western government and an authoritarian regime, individual Westerners usually have a ready-made social entrée into almost any quarter of the globe. With skill, these desires can be manipulated to further a policy, shape attitudes, and persuade potentially hostile populations. A Western propagandist or military advisor normally enjoys an advantage not shared by his rival. English is the universal language of science, technology, aviation, business, and tourism. It is the world's most studied language. Although this fact is of great assistance to British and Americans engaged in military assistance, especially with an indigenous officer class, it is probably more important in gathering information. For example, a Chinese or Russian who only has command of his native tongue is severely limited by the number of local sources of information he has access to in a foreign environment. All things being otherwise equal, the English speaker has the advantage in being informed and in being understood. This Western advantage is particularly important in the realm of low-intensity conflict, where intelligence, advice and assistance, and teaching play vital roles.

In any form of conflict, the steady march of tactical dispersion has produced better chances for a soldier's battlefield survival, but this happy trend may be coming to an end. In 1950, there were only two nations that possessed nuclear weapons and five capable of using chemical weapons. By 1988, a total of nine nations were either known or thought to have nuclear weapons, and fifteen were believed to be capable of chemical warfare. Sadly, the Iran-Iraq War brought the "chemical genie" out of the bottle, and the use of nuclear weapons is also a growing and grim universal possibility.

One thoughtful study projects proven nuclear powers growing from six in the late 1980s to twenty-two known or suspected nuclear powers in 2010. Those countries would include Israel, Saudi Arabia, South Africa,

Argentina, Chile, Pakistan, Egypt, Iran, Iraq, Libya, and possibly Japan. Also, the Republic of Korea, the Republic of China, and the Federal Republic of Germany might be tempted to produce weapons of mass destruction. Additionally, Marxist states that might join the nuclear ranks may include both the Socialist Republic of Vietnam and the Democratic People's Republic of Korea.[28]

There are two conceivable effects of the proliferation of nuclear and chemical weapons. The first effect makes the possibility of both nuclear war and low-intensity conflict more likely while lowering the probability of mid-intensity conflict. An accidental or unplanned use of nuclear weapons is simply more possible with a larger number of nuclear powers than it is with a smaller number. High-intensity conflict therefore becomes more likely. There are, however, inhibiting factors to the incidence rate of mid-intensity conflict. The risks of societal suicide are raised with the diffusion of both nuclear and chemical weapons. Traditional invasion conducted by conventionally organized and equipped armies against powers that may have either or both of these classes of weapons is progressively less of a preferred policy option as such weapons become more widespread. But then, there are always national leaders who are seeking to achieve goals through the use of force, and these new inhibiting factors associated with mid-intensity conflict are apt to make the use of low-intensity conflict more attractive.

Another effect of nuclear and chemical proliferation is further erosion of the global policeman roles of the two 20th Century superpowers, the United States and the Soviet Union. The marginal power advantage of both nations over other states is somewhat reduced with each addition to the list of nuclear and chemical powers. In a very tangible way, the year 2010 will see power more significantly diffused than it is today.

Global Changes

Power sharing in the early 21st Century will be in marked contrast with the old arrangements of the late 20th Century. Rather than a world dominated by two superpowers, there will be five somewhat evenly matched centers of power: the United States, the Soviet Union, Japan, China, and a confederation of West European states. At least four of these powers will have the ability to destroy one another. Whether Japan chooses to become a nuclear power is contingent on a number of unknowns—the possibility of a U.S. military withdrawal from Northeast Asia, a nuclear North and South Korea, and a fundamental reassessment of Japan's role in the world by its leaders and its people. Barring some wholly unforeseen event, Japan will likely forgo a nuclear capability, but it will nonetheless be a world industrial and financial power of considerable dimensions. The five great powers will probably have a rough parity in economic and technological strengths, but the Soviet Union could become the preeminent military power if it somehow solves its ethnic, economic, and alliance problems, and if the West fails to maintain military strength. The bipolar world of the late 20th Century will become merely a subject for historical study. The year 2010 will see a pentapolar world.

There will be a number of obvious issues shaping relationships between the five powers: rivalry for disappearing oil reserves, cooperation or competition in a growing number of Third World conflicts, fears raised by the continued strength of the Soviet armed forces, and the occasional unity of the West in the face of various threats. No issue will be more important than the possible destabilization of the Soviet empire. The relentless separation of Moscow from its artificial shield, the western members of the Warsaw Pact, coupled with the rise of Soviet irredentism and internal ethnic nationalism, could result in a gradual and peaceful diminution of this mighty military power, or it could end in a violent global conflagration.

Most of these changes can be dated from 1975 and an initial phase of apparent communist victories in the Third World. Almost immediately, a role reversal began to occur. By 1985, there were fifteen insurgencies where guerrilla organizations were in strengths of 5,000 or more. The total number of insurgents in those organizations was estimated at over 660,000, and of that number over two-thirds were fighting against Soviet-supported Marxist regimes, a dramatic change from a few years earlier when almost all insurgents were fighting Western-supported governments.[29] Some of the new anti-Marxist insurgent organizations were receiving Western encouragement and supplies. It is perhaps with that sobering point in mind that Mikhail S. Gorbachev revised a long-time policy of Moscow, renouncing "doctrines that endorse the export of revolution."[30] He may have been inviting a similar policy from the West so that the U.S.S.R. could cut its growing foreign assistance outlays to guerrilla-besieged Marxist regimes in the Third World. Gorbachev may also have been looking into a grim future for Moscow, a future with distinct possibilities of armed revolt within the Soviet Union, but Gorbachev could have misread the trend. The movement that began in 1975 could be more accurately described as a worldwide insurgent campaign against authoritarian states, a movement that is independent of Western actions and a bipolar world.

Whatever the case, the incidence of low-intensity conflict is increasing, and the world's special operations forces will consequently see much service. How these conflicts are fought and how special operations forces are employed will depend in some measure on technology. That too is changing, particularly for the insurgent, the terrorist, and the peacekeeper.

Notes

[1] U.S. Central Intelligence Agency, Directorate of Intelligence, *Handbook of Economic Statistics, 1987* (Washington, D.C., 1987), pp. 24–25. Report of the Orga-

nization for Economic Cooperation and Development as reported in the *New York Times,* May 22, 1988, p. F6.

[2] Steve Lohr, "European TV's Vast Growth," *New York Times,* March 16, 1989, p. 1.

[3] Statement of Austrian Chancellor Franz Verzetnitsch as quoted in the *New York Times,* July 24, 1988, p. 8.

[4] Chancellor Helmut Kohl, as quoted by Friedrich Thelen, "Kohl Calls for Army of Europe," *Defense News* (October 24, 1988): 1.

[5] Vice Admiral Michel Merveilleux, Commander, French Ballistic Missile Submarine Fleet, as quoted in *Jane's Defense Weekly* (October 10, 1988): 1050.

[6] Mikhail S. Gorbachev, *Perestroika: New Thinking for Our Country and the World* (New York: Harper and Row, 1987), p. 204.

[7] Major General Edward B. Atkeson, "The New Pact Doctrine: Implications for N.A.T.O.," *Army* (November 1988): 11–14.

[8] British Ambassador to the United States Sir Anthony Acland, "Europe Bears the Defense Burden," *Defense News* (October 24, 1988): 28.

[9] January 1988 poll of registered voters by the Daniel Yankelovich Group as reported in the *New York Times,* February 10, 1988, p. 2.

[10] "Public Summit 1988," conducted by the Public Agenda Foundation and Foreign Policy Development Center of Brown University, as reported in the *New York Times,* March 30, 1988, p. A27. George Gallup, Jr., *Forecast 2000* (New York: William Morrow and Co., 1984), pp. 120–140. The 1989 Roosevelt Center for American Policy Studies poll and the 1989 Americans Talk Security poll, as quoted in the *Christian Science Monitor* (May 3, 1989): 7.

[11] David M. Abshire, *Preventing World War III: A Realistic Grand Strategy* (New York: Harper and Row, 1988), p. 258.

[12] Former Energy Secretary James R. Schlesinger, "Meet OPEC's New Friend: Us," *New York Times,* January 4, 1989, p. A21. Matthew L. Wald, "Why Americans Consume More Energy to Produce Less," *New York Times,* March 12, 1989, p. E6. See also Wald's "An Energy Glut in the Ground Imperils Ecological Hopes," *New York Times,* October 15, 1989, p. E5.

[13] Stuart Auerbach, "America, the Diminished Giant," *Washington Post,* April 15, 1987, p. A1.

[14] Robert A. Fuhruman and Robert M. Gardiner, members of the U.S. Defense Science Board, "Technology Is Our Best Defense," *New York Times,* November 20, 1988, p. F3.

[15] Charles Herzfeld et al., *Technology for National Survival* (Washington, D.C.: U.S. Department of Defense, Commission on Long-Term Strategy, 1988): 14.

[16] Fred C. Iklé et al., *Discriminate Deterrence* (Washington, D.C.: U.S. Government Printing Office [USGPO], 1988), pp. 6–7.

[17] Frank C. Carlucci, *Soviet Power: An Assessment of the Threat* (Washington, D.C.: USGPO, 1988), p. 32.

[18] Ruth Leger Sivard, *World Military and Social Expenditures, 1987–1988,* 12th ed. (New York: World Priorities, 1987), pp. 24–26.

[19] A.F. Mullins, *Born Arming: Development and Military Power in New States* (Stanford, CA: Stanford University Press, 1987), pp. 113–114.

[20] Robert L. O'Connell, *Of Arms and Men: A History of War, Weapons, and Aggression* (New York: Oxford University Press, 1989), p. 306.

[21] Director of Central Intelligence William H. Webster, as quoted in the *New York Times,* February 10, 1989, p. A3.

[22] Clyde H. Farnsworth, "The World Debates the Food Supply," *New York Times,* December 11, 1988, p. E3.

[23] Kevin Kinsella, U.S. Census Bureau, as quoted in the *New York Times,* December 11, 1988, p. 20.

[24] United Nations, *The State of the World's Children, 1989* (New Delhi: U.N. Children's Fund, 1988), as quoted in the *New York Times,* December 20, 1988, p. 1.

[25] U.S. Agency for International Development, "Development and the National Interest into the 21st Century," (Washington, D.C., 1989), pp. 1–25.

[26] Hernando de Soto, *The Other Path: The Invisible Revolution in the Third World* (New York: Harper and Row, 1989), pp. 231–258.

[27] G.D. Kaye et al., *Major Armed Conflicts: A Compendium of Interstate and Intrastate Conflict, 1720–1985* (Ottawa: Department of National Defense, 1985), pp. 128–134.

[28] Charles W. Taylor, *A World 2010: A Decline of Superpower Influence* (Carlisle Barracks, PA: Strategic Studies Institute, 1986), p. 23.

[29] Rod Paschall, "Soviet Counterinsurgency: Past, Present, Future," in Richard H. Schultz et al., ed., *Guerrilla Warfare & Counterinsurgency: U.S.-Soviet Policy in the Third World* (Lexington, MA: Lexington Books, 1989), pp. 151–174.

[30] General Secretary Mikhail S. Gorbachev, speaking before the Cuban National Assembly, April 4, 1989, as quoted in the *Washington Post,* April 5, 1989, p. A1.

CHAPTER THREE

Special Operations and Low-Intensity Conflict Technology

A new era of warfare began to emerge during the last few weeks of the war in Southeast Asia. It was in the waning days of the American involvement, the spring of 1975, a time when little attention was given to a conflict grown too long. What notice was taken was largely focused on the battle in South Vietnam. But the change in the method of waging war was most dramatic in Cambodia. There, American military influence had resulted in a gradual mechanization of the U.S.-supported Army of the Khmer Republic. Propeller-driven T-28s were being used for close air support and 1950s-vintage tracked vehicles, M-113s, were employed as tanks against the Khmer Rouge communist forces. In March 1975, mechanization began to fail.

The increasing battlefield success enjoyed by the Khmer Rouge was partially due to two weapons, the Chinese 85-mm recoilless rifle and the Soviet SA-7 shoulder-fired anti-aircraft missile. Both weapons were easily man transportable and suitable for use by units of the communist forces that were little more than a collection of guerrilla elements. The recoilless rifle was far superior to the Soviet RPG, a hand-held rocket-propelled grenade that was inaccurate, wind-sensitive, and prone to inflict only slight damage on the aluminum-clad M-113 armored personnel carriers. Although the Chinese weapon was heavier and fired from a tripod, it had a good optical sighting device and its powerful warhead literally destroyed the U.S.-supplied tracked vehicles. The armor advantage of the counterinsurgent force had been effectively countered.

The anti-aircraft capability of the Khmer Rouge was more devastating to the fortunes of the Khmer Republic and its American supporters than their enemy's anti-armor capability. The use of the SA-7 quickly changed the battlefield. Air drops to several government-controlled enclaves had to be accomplished from altitudes of 8,500 feet. The result was that many supplies destined for Lon Nol's army drifted into the hands of the Khmer Rouge. Close air support became a thing of the past and armed air reconnaissance ended. Steep landing approaches to airfields were mandatory so as to avoid low flight over territory possibly occupied by the communists, and precautionary flares had to be used to confuse the infrared sensors of the SA-7s. By the spring of 1975, the insurgents had nullified many of the advantages of air power. Warfare, particularly low-intensity conflict, was beginning to change.

Followed to a logical conclusion, effective hand-held weapons can change more than low-intensity conflict. A foot-mobile force that could destroy an opponent's tanks and combat aircraft with regularity would have a number of distinct advantages. These advantages may be realized by 2010 and will probably be demonstrated on a battlefield where a well-equipped insurgent force faces a mechanized army. And, if the war in Afghanistan is any indication of the future, this may become a common occurrence. The change will be far-reaching and will not only enhance the fortunes of insurgents, but affect all types of warfare and military units, including special operations forces, organizations that are expected to perform in all levels of conflict.

Evolution and Revolution

There are so many obstacles to implementing technological change that it is safe to say the outward appearance of most major combat systems of the year 2010 will bear a close resemblance to those of 1990. Even if dramatic technological advances were in hand, the sheer expense and time required to retool industry, test an entirely new system, and adapt it to military and naval forces would put the fielding of such systems out into the 2020 to 2030 time frame. A fully funded weapons program, free of the normal obstacles and false starts, takes the U.S. Department of Defense research and development process twelve years from inception to operating capability.[1] It is highly probable that the tanks, fighter aircraft, submarines, and helicopters of 2010 will look very much like those of 1990. In fact, many 1960-era versions will still be on the world's battlefields and in the inventories of 21st Century armed forces. The Soviet T-62 tank, the American M-113 armored personnel carrier, and the British Chieftain tank will be much in evidence within the forces of Third World states. Even more modernized forces, such as those of the United States, will carry some Vietnam-era equipment. The CH-47 helicopter, for example, will probably still be an American standard in 2010. The Soviet 130-mm and the U.S. 175-mm, 8-inch, and 155-mm artillery pieces will likely remain as the mainstays of the modern forces of the year 2010, as will the Soviet AK-47 and American M-16 rifles. Not only are retooling, test, and adaptation impediments to modernization, so too is consideration of costly sunk investments in a current inventory.[2]

Yet, despite outward appearances, there is apt to be considerable change in even the older combat systems. An American 155-mm howitzer in 2010 may look similar to the version used by U.S. artillerymen in World War I, but the projectile it fires is likely to be radically different. The base of that projectile will probably be chemically treated so that gases are created by air friction in flight, filling the drag-producing vacuum characteristic of older munitions. The result could be as much as a 30 percent increase in range. And modernization will not stop there. The new projectile will not likely explode on impact, as it will have already released submunitions over its target. The submunitions may be briefly suspended over enemy forces by small parachutes while their millimeter wavelength or infrared sensors locate emissions from objects such as combat vehicles. Parachutes will be released and small rockets will ignite, driving the submunitions to the top of their individual targets. If the intended target is a group of infantrymen, the submunitions are likely to be small, camouflaged bag mines that arm themselves on impact with the ground.[3]

Appearances will also be deceiving for the venerable, Vietnam-era CH-47 helicopter. By the early 1990s, American special operations forces are planning to operate between 16 and 30 of the Chinooks, but unlike the earlier versions, these will be air refuelable and equipped with terrain-avoidance radar. Additionally, the pilots will have a global positioning navigation system, as well as a forward-looking infrared capability for night operations. The MH-47E will be able to transport 44 troops within a 300 nautical-mile range and will have extended-range fuel tanks for a smaller combat load.

These changes are evolutionary in nature, yet they greatly expand combat capabilities. Military planning factors will be under constant revision during the 1990s just to accommodate the myriad product improvements that are entering or nearing production. But because these innovations are almost invariably cheaper than programs aimed at creating wholly new systems, they are obstacles to fresh technological approaches.

However, there are several innovations that may be on the battlefields of 2010 that are revolutionary. Low-observable, or stealth, technology aircraft—both manned and unmanned—might account for about 10 to 20 percent of the airframes of modernized air forces. Hand-held meteor-burst communications sets will permit the transmission of short messages over distances of up to 700 kilometers. Unlike current long-range communications technology, an opponent will have great difficulty in locating the sender of meteor-burst transmissions. Both meteor-burst communications and low-observable technology will revolutionize clandestine operations, the domain of special operations forces.

Technological Requirements for Special Operations Forces

Although there are marked differences in the three special operations force models—U.S., Soviet, and British—there is some common ground. All special operations forces share the need to reach the rear areas of their enemies, sustain themselves in those dangerous regions, and communicate in safety. Additionally, these forces require light, man-portable weapons to inflict damage on their opponent's vital rear area facilities—airfields, road networks, and headquarters. They also require a means of returning to their own territory and, even though they will normally avoid contact with an opponent's combat formations, they need self-protection weapons.

All three types of organizations have the mission to accomplish reconnaissance deep in enemy rear areas. If a special operations unit is only charged with this function, it is configured in a slightly different way and must avoid any contact with enemy forces or the indigenous population. Unlike an organization that is tasked with supporting insurgents, it cannot rely on local sources for food and must either bring its own rations or have them supplied. It should not forage for food, as such activity attracts attention. The materiel requirement here calls for lightweight, compact rations that will sustain the unit for extended periods.

Since American special operations forces are a bit more varied and complex, their technological needs range further. For example, U.S. psychological operations elements need the means to communicate with enemy forces and populations. U.S. civil affairs units require information management systems that deal with host nation and enemy governmental organizations, facilities, and personalities, as well as with commercial, industrial, and agricultural systems.

Special operations forces are normally tasked to provide a counterterrorist capability. In this field, materiel requirements include highly accurate sniper weapons and marksmanship training systems. For hostage barricade situations, the counterterrorist force must have effective breaching devices that can quickly open doors, windows, and even walls with minimum hazard to the occupants. This type of activity also calls for comprehensive computer systems that can quickly retrieve the methods of operation of a particular terrorist organization, its probability of killing hostages, vulnerabilities, history of using explosives, frequency of activity, and international connections.

Generally, the more sophisticated materiel for special operations forces is required for high- and mid-intensity conflict. Rudimentary or even obsolete aircraft will often suffice for penetrating the air defense system of a Third World nation in low-intensity conflict, but more capable airframes are needed to get through a modernized air defense arrangement that can be found in mid- and high-intensity conflicts.

Technology and Low-Intensity Conflict

Many of the materiel needs for an insurgent organization are identical to those of a special operations force, particularly if the insurgent enjoys external support. A means to infiltrate men and supplies into the battle area from a secure, external base is required, and hence, there is a demand for penetration aircraft. Both insurgents and special operations forces need light, man-portable communications equipment, radios that transmit signals difficult for an opponent to intercept and decipher. Similarly, compact anti-tank and anti-aircraft weapons systems, as well as demolitions material, are useful to both guerrillas and special operations units. Easily emplaced mines that are both pressure initiated and capable of remote activation are required. Insurgent organizations and special operations forces need handily packaged medical supplies, and they should be clothed in a fashion that promotes concealment, health, and speed of movement.

On the other hand, the counterinsurgent requires a rapid, well-armed air assault capability, area anti-personnel weapons, and sensor systems that can detect human movement in remote regions. There is also a counterinsurgency requirement for a fully integrated, national-level intelligence collection, analysis, and dissemination system. The counterinsurgent has to keep communications routes open, because a prime tactic of the insurgent is progressive isolation of selected areas of the country. That produces the requirement for fast-moving armor formations that dominate road nets and have the capability to make repairs quickly to damaged bridges and other road cuts. Vehicles for such formations should be resistant to the effects of mines. The counterinsurgent should also have attack aircraft suitable for employment against guerrilla formations. They should be built for accurate weapons delivery, nighttime operations, and extended loiter times. Counterinsurgency forces must also have a highly responsive means of intercepting, locating, and decoding an insurgent's radio communications.

Few nations can afford two separate force structures, one for mid-intensity conflict and one configured for counterinsurgency tasks. Counterinsurgency forces are therefore usually created by modifying an existing regular force. The same is true of peacekeeping or peacemaking forces. A peacemaking force conducting stability operations should simply have the materiel and technological edge required to subdue an opponent, and its force structure is therefore contingent on that of a potential enemy. The peacekeeping force usually performs a simple observer role but, if it is to be effective, it must identify threats and often secure a border zone.

Terrorism and Technology

Early 21st Century state-supported terrorism may present problems that will be seemingly impossible to resolve. The government-sponsored terrorist will have a number of potent weapons at hand. Small, easily concealed anti-aircraft weapons, radioactive and biological contaminants, and chemical and nuclear devices capable of clandestine emplacement will probably be available. These types of large-scale casualty-producing threats will undoubtedly result in a greater proclivity to use cross-border raids on the part of the offended parties.

Terrorist groups that engage in perhaps the oldest form of terrorism, assassination, do not necessarily require state support. These organizations will probably have access to two types of weapons that will ensure a higher success rate in their operations: laser sniper systems and improved lightweight anti-aircraft weapons.

Accuracy with a conventional sniper rifle requires a precision weapon, proximity to the target, careful calculation of wind conditions, and a professionally trained marksman. Since protection services available to governments are normally competent enough to secure routes of passage for important personalities, a conventionally armed assassin may have difficulty in gaining proximity to the intended victim. On the other hand, a laser weapon is not as range dependent as a conventional rifle, and it is unaffected by wind conditions. It also has the advantage in "spotting" the target so that the sniper can mark his prey before he initiates a lethal pulsed laser.

Assassination will regrettably become an easier task by 2010. Not only will laser weapons complicate the duties of those charged with protecting political leaders, but the proliferation of shoulder-fired anti-aircraft weapons will make air transport of chiefs of state and high-ranking officials a far more hazardous activity. The physical distance between leaders and the led will probably be lengthened, and a leader's access to his followers will be diminished.

The rescue of hostages may be a bit easier by reason of continued growth in the sheer numbers of counterterrorist forces and the gradual development of improved rescue techniques. The hostage barricade scenario is not likely to enjoy the popularity with terrorist organizations that it once did. Essentially, it is a suicidal activity, and there is growing general agreement among governments that bargaining with terrorist groups only begets more terrorism. Additionally, as in the case of state-supported terrorism, retaliatory raids are becoming an internationally accepted response.

Retaliatory raids fall into the category of active counterterrorist measures, but there is also likely to be considerable technological progress in the field of passive defense, or anti-terrorism. Success against state-supported terrorism is largely dependent on internationally shared intelligence products, multistate cooperation in penetrating terrorist organizations, and mutual efforts to identify, track, and intercept individual terrorists.

By 2010, it will be electronically possible to read, memorize, and transmit a handprint. It will also be possible to provide quickly an individual's identity through blood or hair samples. One or several of these techniques may be used to replace the current photo visa for international travel, a document that is often illegally duplicated. Terrorism alone will probably not bring about such an innovation. Diseases such as AIDS, international cooperation in combating illegal narcotics activity, and counterespionage efforts all constitute pressures for more stringent control of international travel. These pressures translate into a growing need for a more certain identification of travelers.

The prime advantage of precise physiological identification is that it requires the potential traveler to appear at the embassy or consulate of the nation he chooses for a destination. He exposes himself to other means of identification gathered through intelligence resources at a location regarded as foreign soil. The measure would probably not intercept a newly recruited terrorist, but would serve as a deterrent for the known terrorist. There are a number of technological innovations that are apt to be used against the terrorist of the early 21st Century but, on balance, the march of technology favors the terrorist.

Air Penetration Vehicles

In the field of aviation technology, the trends appear to favor insurgents, at least those who may have the advantage of external support by an advanced military power. This is not due to stealth or, more accurately, low-observable technology alone, but also to improved navigation aids and remotely piloted vehicles (RPVs).

Already, there is an expensive and highly complex race between two competing technologies, one devoted to the creation of low-observable aircraft, the other aimed at achieving means of identifying and intercepting air penetrators. Penetration aircraft, including cruise missiles, are being constructed of radar-absorbent materials and shaped so as to deflect pulsed electromagnetic energy away from radar receivers. In some cases, "smart skin" is used, consisting of embedded sensors and electronic "adjusters" that identify and distort intercepted enemy radar beams. The end result is not likely to be electronic invisibility for penetrating aircraft, but greatly reduced radar cross-sections and an attendant compounding of the defender's difficulties.

These technologies are causing air defense systems to be heavily modified and, in some cases, wholly revised at great expense. A typical approach for the defender is to invest in a large number of ground-based, passive radar receivers in the hope that his essential active transmitters will survive anti-radar weapons systems launched by the attacker. But low-observable aircraft, in most cases fresh technological designs, are normally configured for low-level penetration paths. The defender, therefore, augments his ground-based air defense system by making more outlays for airborne, high-power, low-frequency radar systems. He must also integrate look-down, shoot-down interceptor aircraft and invest in detection systems that identify heat emissions from penetration aircraft.

The likely result is that the defender must greatly outspend the attacker. The creator of the low-observable penetrators has to pay considerable sums to develop the technology, but once that is paid, his investment in personnel to maintain his penetration systems is minimal. On the other hand, the defender has to create both ground and airborne detection systems, as well as the interception means. Additionally, the defender faces great costs in the maintenance of his extensive defense complex, since the defense system is of little use unless it is operated at near around-the-clock schedules. The attacker, of course, chooses his own time and place to penetrate the defender's airspace. Both costs may be high, but the attacker has the cost advantage.

Another forthcoming technological advance that could benefit the insurgent and special operations unit is the enhanced cruise missile. Air infiltration of men and materiel into a hostile area can be extremely hazardous, and such enterprises may often be politically unacceptable on the basis of the anticipated attrition of men and machines. Then, too, the possible disclosure of the sponsoring nation's identity through capture of a downed aircrew in hostile territory provides another reason for a decision maker's reluctance. By far the largest number of sorties for such operations are supply missions, not personnel infiltration sorties. By 2010, the risks associated with air supply tasks can be substantially reduced by using RPVs for the logistics support role. Insurgency support can thus become less risky for the politician.

Although a number of nations are developing and producing a variety of cruise missiles, perhaps future capabilities are best exemplified by the American Tomahawk. The land-attack version of this missile has an air-breathing turbofan engine, travels at altitudes that can vary from just over the earth's surface to 10,000 feet, and delivers a 1,000-pound warhead 700 miles from the point of launch. Navigation is provided by a terrain-matching radar system, but newer models are being augmented with a scene-matching correlator that uses digital photos matched against negatives stored in the on-board computer.[4]

The Tomahawk is extremely accurate and is undergoing constant enhancement. Some idea of its versatility was revealed in 1988 when a test missile destroyed three separately located targets by dropping submunitions on each of them. The missile then dived into its fourth and final target. A newer version will incorporate an improved turbofan, increasing thrust by 19 percent and decreasing fuel consumption by 3 percent. The numbers of U.S. Navy launch platforms for the Tomahawk will be increased from 15 to about 200 by the mid-1990s.[5] By the late 1990s, this missile is apt to achieve even greater payloads and range through the use of ceramic engine components and external propfans. It is likely to be even more difficult for a defender to identify, a result of the shift from a radar navigation system to one with fewer electronic emissions.

For an insurgency support role, missiles such as the Tomahawk must be modified to substitute a cargo bay for the current submunition stores and warhead area. The missile would carry weapons, ammunition, demolitions material, rations, and medical supplies, all packaged in man-transportable bundles attached to radio-controlled parafoil or ram-air parachutes. Remote-control parachute systems are well proven and can be guided from the ground by a recovery team after the supplies are released over the target area. The missiles themselves could then be programmed to return to their launch area or strike a target after their supply tasks are completed. Use of this type of resupply system would greatly decrease the risks and costs associated with support of insurgents. The system would also be highly useful to special operations forces conducting extended-duration missions in enemy rear areas under the conditions of mid- or high-intensity conflict.

The Problem of Communications

Perhaps the greatest period of vulnerability for an insurgent or a special operations unit is during the time it communicates with an external control element or headquarters. Enemy rear area security or counterinsurgency forces have many advantages in intercepting and locating a hostile transmitter. In these regions, government forces can afford to have fixed intercept stations with large antenna arrays, stations that continuously sweep the electromagnetic spectrum and are capable of quickly initiating direction-finding techniques to pinpoint any suspicious transmission.

Even if clandestine transmissions are well enciphered, a security service or counterinsurgency element can learn much from traffic analysis and the geographical pattern of successive transmission sites. Traditionally, communications has represented a prime weakness for both the insurgent and the special operations force. Standard procedures for such elements include infrequent transmissions, short messages, use of the lowest possible power settings, and frequent movement of the transmission site. All of this limits the utility of special operations forces operating in an enemy's rear and creates vulnerabilities for the insurgent. But these limitations can be at least partially offset by 2010.

Up to the 1980s, man-pack, long-range communications systems normally used the high-frequency (HF) portion of the electromagnetic spectrum, frequencies from 3 to 30 megahertz. Use of this part of the spectrum takes advantage of sky wave propagation to transmit messages for hundreds of miles. Transmissions of this nature are bounced off the ionosphere, 50 to 500 kilometers above the earth's surface. Such transmissions are easy for a security force to intercept and locate.[6] Clandestine radio operators could gain some protection by the use of burst transmissions, the technique of encoding and storing a message ultimately sent at high speed, requiring only a minute or so of transmission time. However, by the 1970s, most security services were using computers to sweep the HF portion of the electromagnetic spectrum, computers that were programmed automatically to record suspicious transmissions. Clandestine communicators using the high-frequency spectrum will receive an additional blow in the early 1990s. A skillful operator can make his transmission appear as a maritime message if his sending site is near the ocean. However, in 1993, maritime Morse code will be discontinued in favor of automatic communications systems.[7] The Morse code operator in the 1990s will attract attention. For the clandestine radio operator, HF transmissions, burst or not, will have become a dangerous means of communications.

In the early 1980s, frequency hopping sets became available and it was then possible to establish long-range radio nets of clandestine radios and their controlling base stations that transmitted and received messages on multiple, programmed frequencies. Security force intercept stations could not readily piece together an entire message from any one of the clandestine transmitters, and decryption of these frequency hopping messages became extremely difficult. However, the transmitter and the base station could still be located. At the same time, in the early 1980s, man-pack satellite radios became available; a clandestine transceiver of only thirty pounds could send and receive a voice message over thousands of miles with great reliability. These radios used low power settings, making them difficult for a ground intercept station to hear and locate. All told, the special operations and insurgent communicators were gaining on their traditional enemies.

However, in the realm of mid- or high-intensity conflict, communications satellites are likely to be jammed or destroyed. Even if they were functional, allied insurgents and friendly special operations forces might not merit the priority to use communications satellites, an essential consideration in the choice of communications systems. But, once again technology provided the clandestine radio operator an advantage.

In the 1960s, the U.S. Navy conducted extensive research into the phenomenon of meteor-burst communications. It was found that the lower part of the ionosphere is constantly marked by tiny objects that produce traces of material whose density is sufficient to reflect electromagnetic energy in the very-high-frequency (VHF) range. Meteor strikes occur often enough to ensure a very brief communications path every three or four minutes between any two points on the earth that are approximately 100 to 700 kilometers apart. The technology is currently in use in the United States and has been functional since the 1970s. Master base stations in the western part of the U.S. continuously transmit to hundreds of remote, unmanned outstations. The outstations have sensors that monitor wind conditions, stream velocities, rainfall, barometric pressure, and temperature. Powered by solar panels, the small, automatic transmitters at the outstations instantly send their stored data when the transmitter at the master base station finds the "meteor scar" in the ionosphere that will provide a path to the appropriate remote transmitter site.

Meteor-burst radio techniques represent a major advance for clandestine communications. The intercepting receiver of an internal security service would have to be within a rather small geographic ellipse, approximately 12 by 30 kilometers in size, to locate a transmitting clandestine radio. The master base station must transmit continuously, but it can be many miles away, in secure territory. The master station can also consist of a small, roving ground element linked to an airborne relay station, ideally a long loiter RPV. This type of clandestine communications system offers an unprecedented degree of security for the insurgent or the special operations unit operating behind enemy forces. Although the cat and mouse game between the clandestine radio operator and the world's security forces will continue apace, the odds are increasingly favoring the harried man who works in his enemy's backyard.

The Fine Art of Modifying Nature

The ability to cause rain, fog, overcast conditions, massive forest fires, or defoliated jungle areas is not apt to see much change by 2010. Widespread health and environmental concerns in the West, the origin of such technology, will probably impede further development of existing techniques. However, this type of activity will probably be employed, largely by Third World nations.

While these techniques are of doubtful utility in mid- or high-intensity conflict and have minimal advantages for the operations of conventional forces, they do offer some promise in the fields of low-intensity conflict and special operations. When it was noticed that rain had a suppressive effect on street demonstrations, the U.S. Military Assistance Command in Vietnam used silver iodine crystals to cause showers during the Buddhist unrest of 1963. Commando teams were often infiltrated into North Vietnam under cover of artificially created overcast, usually stimulated by airborne spraying of potassium and chloride aerosols.[8] These man-induced conditions are normally only possible in southern hemisphere tropical climates, and even there they are largely restricted to provoking what might happen in any event. But, in critical and high-risk situations, these techniques could be helpful to both the counterinsurgent and the special operations unit and it should be expected that they will be used.

Deliberately set forest fires in guerrilla-controlled areas were also attempted during the Vietnam War, but the results were less than dramatic. Effective burning is highly dependent on nature, requiring a surface layer of dead wood, dry leaves, and dry grass; protracted low humidity; and sparse cloud cover. Additionally, the ignition pattern is critical to success. Prior chemical defoliation can enhance the results, but a humid rain forest is a poor candidate for such projects.[9] It is interesting to note that the forest fire most damaging to the fortunes of the Viet Cong was one that was not set by either U.S. forces or those of the Saigon government—it was started by irate Vietnamese peasants in retaliation for communist guerrilla-imposed taxes.

There is one facet of nature modification that bears careful consideration, terrorist acts to stimulate earthquakes. Geographic fault lines can be disturbed through the use of explosives or the pressure injection of fluids. Again, it is a condition that may occur in any event, but that point of logic is apt to be lost in a wave of public hysteria, precisely what the terrorist who seeks recognition for his cause wants to bring about. Those officials concerned with anti-terrorist responsibilities must consider daily surveillance of fault lines, particularly those adjoining major population centers.

Naval Special Operations and Technology

The fast-paced race between low-observable and surveillance technology is not restricted to the air, but extends to the oceans as well. In this region, there is little if any lead for the surface attacker, the prime user of low observables, primarily because radar is far more effective where there is no phenomenon of terrain masking. The 21st Century should therefore see an increased use of subsurface operations, a region that poses the most severe surveillance problems. It can be expected that mid- and high-intensity conflicts at sea will be heavily influenced by submarine-launched cruise missiles with 1,500 to 3,000 mile ranges, and those subsurface craft will be equipped with anti-aircraft missiles for their own defense.[10]

Naval special operations forces such as the Soviet Naval Spetsnaz, U.S. SEALs, and the British SBS have a somewhat limited utility in low-intensity conflict, but they are likely to have an increasing role in mid- and high-intensity conflict. Both the United States and the Soviet Union are planning to exploit a vulnerability of naval forces, the dependence on fixed ports. Barring attacks by intercontinental ballistic missiles, key naval facilities are usually well protected from air attack, whether attack by cruise missiles or conventional aviation. Additionally, hostile submarines would have difficulty in gaining proximity to these ports. However, they can be approached by undersea craft, and submarine-launched special operations forces may thus be able to infiltrate the defensive perimeters. The special operations forces of both nations are developing capabilities to attack such targets.

It is believed that the Soviet Navy has heavily invested in small underwater reconnaissance and infiltration craft for its Spetsnaz forces and that these craft have been successfully tested against the coastal defenses of neutral Sweden. It is known that the U.S. Navy plans a considerable program of this nature in the 1990s. The American SEALs currently have fifteen submarine-launched infiltration craft in both two-man and six-man versions. The two-man craft is "wet," the underwater swimmers riding the near silent craft externally. The six-man version has an internal dry compartment for the infiltration team.[11] These craft are programmed to be increased in numbers and improved, but the major American effort to enhance this specialized capability is the planned conversion of a number of nuclear submarines for SEAL use, providing "mother" boats for launching the smaller, submerged coastal and port infiltration craft.

The Counterinsurgent's Area Weapons

As in the case of defoliants, forest fires, and weather modification efforts, Third World nations are likely to use chemical and other types of area weapons in future counterinsurgency campaigns. Already, North Vietnam has been suspected of employing chemical weapons against Hmong tribesmen in Laos, and it is certain that Iraq has used gas against its own Kurdish tribesmen. While it is doubtful that new chemical agents or other material will be specifically developed for this purpose, innovation in the means of delivery is apt to increase the effectiveness of such weapons.

Western nations have been creating the means to destroy Warsaw Pact airfields, and this technology can be adapted for another purpose. For example, the BL-755 cluster bomb dispenser, capable of delivering 49 runway denial mines, provides coverage of an 80 by 35 meter area. A single aircraft can deliver 18,500 submunitions, covering a one square kilometer target area.[12] The British-developed JP-232 delivers several hundred bomblets that penetrate concrete. These bomblets could be exchanged for fuel-type submunitions that create highly flammable aerosols on impact, submunitions that are triggered for delayed ignition. Fuel-air weapons were used during the Vietnam War to clear helicopter landing areas in dense jungle. These early, crude versions were actually huge tanks containing a propane and TNT slurry.[13] Used against personnel, these weapons create a wave of massive overpressure, literally killing by the acre. In the frustrating task of hunting down elusive guerrilla units, it is likely that area, anti-personnel weapons such as chemical, improved fuel-air munitions, and mines will be efficiently dispensed over broad expanses of difficult terrain.

Sustainment in Special Operations and Low-Intensity Conflict

The ability or inability of a small unit to sustain itself without external support for extended periods often spells success or failure in either low-intensity conflict or special operations. No one expects a tank force to succeed without a massive logistical organization, thousands of gallons of fuel, and a horde of mechanics and spare parts following in train. The same is true of air power and naval operations. There must be the ever-ready and well-appointed air base, floating or otherwise, and there must be the naval port, handy supply and repair shops, and stocks of spare parts. Special operations and low-intensity conflict are different. These two activities demand efforts by small, independent, but coordinated groups, groups that can exist, operate, and succeed with little aid from outside the battle zone.

For example, a guerrilla force that is dependent on food being brought to it by an external sponsor has little chance against the competent counterinsurgent. Logistical support on a frequent basis and air delivery flights will, in the end, reveal the guerrilla's location. Even if the guerrilla cannot be caught, the logistical support itself can be denied. The same imperatives apply to special operations units involved in reconnaissance missions deep in an enemy's rear. In both fields, special operations and low-intensity conflict, feeding, clothing, and sustaining small units for protracted periods with minimal external support are often goals of utmost importance. Here, technology can help and, in fact, it already has.

One of the great logistical successes of warfare occurred during the Vietnam War with the rapid development of lightweight rations for American long-range reconnaissance elements. The usual task for these special operations units during the early phases of the war was to locate their enemies' logistical arteries extending through the mountainous jungle chain from North Vietnam, then through Laos and into South Vietnam. A good performance required extensive walking through a target area, stealth in movement, rapid escape from enemy security forces posted along the trails, and protracted surveillance of the assigned region. Very quickly, it was found that the weight of food required to sustain such units was an important limitation. Foraging for local sources of food severely detracted from the mission at hand and to attempt air resupply would have risked compromise of these reconnaissance elements. American Special Forces teams, therefore, had to carry everything they ate from the very start of their missions. Having only the standard rations at the beginning of the war, three to four pounds of food were required for each man, each day. Adding the weight of radios, batteries, weapons, ammunition, and minimal medical supplies, and considering the need to be able to run quickly, the duration of each mission was obviously limited.

At the same time, the Americans were studying their prey, the Vietnamese porters were carrying supplies to their comrades down the long trail south. It was discovered that thirty ton-miles of throughput were being achieved by ninety to a hundred porters, each consuming an average of 2.7 pounds of food per day.[14] Within eighteen months, the U.S. Army's Natick Laboratories developed freeze-dried and irradiated food technologies and fielded rations of 3,000 calories in an unreconstituted weight of only two pounds. The new Long-Range Reconnaissance Patrol Rations immediately extended the employment periods for U.S. special operations teams along the Ho Chi Minh Trail and provided a further capability that, for the most part, was rarely used during the Vietnam War. These compact ration packets, when hidden in caches, could long survive the extremes of almost any climate, a rare characteristic for food supplies.

Reconstitution of these rations does, of course, require a water source, but water is needed in any event. Further development of this important reconstitution technology in the 1980s resulted in additional improvements. It was found that in a moderate climate, a two-hundred-pound soldier performing relatively sedentary activity such as a surveillance task needs only about 2,100 calories for sustained operations. That ration has been produced in a package of less than one pound and in bulk of less than 45 cubic inches.[15] Few technological developments are more vital for enhancing the performance of the special operations soldier and the guerrilla.

Insurgents, operating in a rural environment, and special operations teams employed in deep surveillance or reconnaissance tasks should also enjoy marked improvements during the late 1990s and early 21st Century in body armor, load-bearing equipment, clothing, and footwear. These seemingly mundane enhancements are of great importance. Together, they contribute to a wholly new level of survivability, health, and mobility for small, widely scattered units that are employed in great depth.

By the mid-1990s, body armor and scientifically designed load-bearing equipment should finally be integrated into vests of great resilience that evenly distribute weight about the torso. Body armor has steadily improved throughout the 1980s, based on Swiss tight-weave manufacturing techniques that produce pliable, relatively thin vests of multi-layered synthetic material. These vests have substantial bullet and fragment stopping power and are far lighter than their Vietnam-era predecessors, vests that were often discarded in view of their weight and lack of wearing comfort. At the same time, a thriving hiking industry in the United States has produced body-conformable packs that allow the carrier to maintain balance and transport seventy-pound loads with surprising ease. What remains is to design the placement of standard items such as ammunition, knives, compasses, canteens, and grenades on the vest, calculate the bullet-stopping power of these items, and eliminate an appropriate amount of the vest material where the item's pouch is placed. The result could be a combination load-bearing and armor system far superior to the bulky, uncomfortable, and heavy separate components of the 1980s. Mobility and survivability of the foot soldier would be measurably enhanced.

Some of the same types of technology and commercial activity are now producing vastly improved footwear and weather-resistant field clothing. The jogging fad of the 1970s produced significant advances in podiatry and the manufacture of lightweight athletic shoes that absorb shock, enhance speed, and are capable of long wear. By the 1980s, this technology had been applied to the booming hiking industry and resulted in lightweight boots that dry quickly, "wick" casualty-producing moisture from the foot, and provide comfortable support for long marches.[16] Finely woven synthetic materials have also been produced for outdoor wear. Microfiber insulation, composed of polypropylene and polyester, can provide almost twice the insulation per centimeter of thickness as down, the old industry standard. These materials are capable of retaining body heat with ultra-thin insulation while allowing enough air exchange to draw off perspiration. They repel wind and rain, but are quite lightweight and resistant to wear and tear.

The cumulative effect of these advances in food, body armor, clothing, and footwear technology will be to extend greatly the endurance of small, footmobile units and improve their levels of survivability and performance. By the turn of the century, survivability and endurance should see a near 100 percent enhancement over similar levels of the 1960s. Few fields of military technology will see such dramatic improvement. Taken together, these changes amount to a revolution in low-intensity conflict and special operations.

Psychological Operations and the New Technology

There is no indication that there will be any sort of a psychological operations revolution based on technology, at least in the fields of print and radio broadcast media. But it can be expected that a novel technique of television may be in use by 2010. In this realm, however, it is the message that counts and, if the message is not right, no amount of technology can make it effective.

It is likely that there will be a growing use of RPVs for psychological operations. In part, this will stem from the understandable reluctance to risk airframes and aircrews in the increasingly hostile air environment of the future. There is little likelihood that psychological operations will count for much in a conflict of high intensity, but it can be expected that RPV-delivered leaflets will be used in mid-intensity conflicts, particularly those of extended duration when troop battle fatigue and public weariness become factors.

The novel innovation in future psychological operations has to do with television insertion, a technique that involves the use of an RPV. Television has already become a dominant medium for public information and entertainment, and it is not uncommon to see soldiers combating boredom by viewing small portable sets in the most unlikely places. Additionally, the common videocassette recorder (VCR) has already made a political impact. The KGB is much concerned with illegal TV tapes in the Soviet Union and has even created a word for copying and distributing disturbing products: *magnitzdat*, or tape publishing.[17] The pervasive use of television is a trend that can be expected to continue throughout the world, and that trend must absorb the attention of those engaged in psychological operations. One characteristic of television that makes it an appealing medium for psychological operations is that a television broadcast takes up 250 channels of voice communications and is therefore highly susceptible to jamming, interference, or manipulation.

By the late 1990s, a growing percentage of Western television sets will be receiving high-definition broadcasts, transmissions that deliver pictures of great clarity. By the turn of the century, high-definition television is expected to be a global enterprise, and by 2010, a large portion of the world's television sets are expected to be capable of receiving these broadcasts. The propaganda significance of high-resolution television is that signals for these broadcasts are likely to be of a high-speed nature, speeds that facilitate clandestine insertion of subliminal messages.

The human eye and thought process is capable of receiving and subconsciously understanding a visual message only .001 seconds in duration. This subliminal message is added to the brain's memory without conscious knowledge of the message's origin. Using a standard television broadcast, the viewer is apt to be aware of the inserted message, but in viewing high-definition television there is a high probability that the audience will fail to connect the message with the screen. Viewers mentally register a thought, one they might have rejected given conscious consideration. While it is doubtful that the inserted thought will modify behavior by itself, it does provide a foundation for successive messages built around a theme.[18] Delivery of subliminal messages is likely to be by RPVs capable of brief high-energy pulse transmissions designed to surreptitiously integrate with and ride existing television transmissions. Fortunately, the West, with the steadily growing use of cable TV systems, is less susceptible to this type of attack than other parts of the world that continue to depend on air broadcasts.

The Case of RPVs

The stultifying bias of those who favor manned war-planes may prevail over the advocates of remotely piloted vehicles into the 21st Century. However, in the event RPVs are seriously developed, they will have a considerable impact on special operations and low-intensity conflict. Some mention has already been made of the use of robot cruise missiles in a resupply role for special operations forces and insurgent groups, and in a potential psychological operations role. In addition, there are a wide range of other tasks that RPVs may perform, such as reconnaissance, communications relay, deception, and target attack.

The gradual ascendency of air defense systems over penetration aircraft should produce a greater use of RPVs, particularly for reconnaissance over enemy territory. Man's physiological characteristics are already performance-limiting factors for some modern aircraft, and it is possible to build RPV airframes that exceed performance levels of manned aircraft. RPV photo reconnaissance missions can be accomplished in a difficult threat environment by using an aircraft that can sense air defense missiles and outmaneuver them by performing dives, turns, and climbs that a pilot could not endure. Such RPVs may be necessary, even in a low-intensity conflict, due to the programmed mid-1990s emergence of Soviet SA-16s and the newest version of the U.S.-made Stinger, second- and third-generation shoulder-fired anti-aircraft weapons. At higher altitudes, out of the range of these hand-held air defense missiles, long loiter time, quiet RPVs will probably be used in a counterinsurgency role carrying high-resolution, zoom lens television cameras and sensors designed to detect insurgent activity.

The RPV has a particularly useful role in both special operations and low-intensity conflict when it is used as a communications relay platform. For the counterinsurgent, a high-altitude, long-flight-duration communications RPV provides an excellent opportunity for counterguerrilla patrols to employ very small and light radios, an important foot mobility factor. For the support of insurgents, these airframes can be used over friendly or neutral territory in order to capture the insurgent's incoming messages and transmit the master base station's traffic without giving away the location of the latter. A special operations element needs the same type of capability. Such an airframe is already available in the Teledyne Ryan Spirit, an RPV that can achieve 50,000 feet in altitude and 80 hours of loiter time carrying a 300-pound payload.[19] The Spirit has an 85-foot wingspan, a near ideal surface length for the reception and transmission of meteor-burst communications.

As a deception platform, the RPV is highly useful to the sponsor of insurgents and to the special operations unit. It is useful to test and define a high-threat air defense system by using an RPV flying an offset insertion pattern prior to personnel air infiltration missions. During the infiltration mission, it may be essential to launch several RPVs to confuse and saturate a segment of the defender's surveillance and interception net. These RPVs should have a radar augmentation system that emits a variable radar cross-section. They should also be constructed of radar-absorbing composites or have the ability to distort electromagnetic energy so that they can return safely after their task is completed.

There is also a special operations and low-intensity conflict role for the "one-way," or killer, RPV. The primary mission for such airframes should be the destruction of tanks. Rapid advances in armor technology make it imperative to have a weapon that is capable beyond doubt of penetrating modern armor. Such a weapon will necessarily have either a large warhead or a smaller, sophisticated one. By 2010, it should be possible to manufacture an RPV that is autonomous, or at least semi-autonomous—that is, the vehicle "thinks" for itself. These unmanned aerial vehicles (UAVs) must have sensors that not only locate targets, but have the capability to identify hostile threats to themselves. They must be programmed to destroy one and avoid the other without an operator's intervention.[20] It is certain that a defender operating against UAVs will bend every effort to counter them, so it will become important for the attacker to limit the duration of these killer UAV flights. A method of doing that is to place such RPVs within a special operations structure so that teams operating within enemy territory physically locate appropriate targets and provide general target areas or even terminal guidance.

The Problems of Tanks and Airplanes

Although the tank had been used as a weapon of war since 1916, and airplanes had seen action as early as 1911, it can be said that the Battle of Cambrai in November 1917 marked the beginning of mechanized warfare. There, large numbers of tanks, motor trucks, and airplanes were fully integrated in a coordinated plan of attack. No one can be absolutely certain when the era of mechanized warfare will end, but surely it will—all eras eventually pass. What is certain is that tanks and combat aircraft are becoming increasingly vulnerable to each successive generation of anti-tank and anti-aircraft weapons, some of which are easily man-portable and shoulder-fired.

Armor designers of the 1970s and 1980s progressively resorted to more radical solutions in keeping the tank as a viable weapon system. As the sensors, guidance systems, and warheads of anti-tank weapons steadily improved, simple rolled homogeneous steel armor gave way to the British Chobham design of alternate layers of aluminum, steel, ceramics, and fabric. That, in turn, had to be augmented by reactive armor, boxes of explosives attached to the tank's hull that disrupt the penetration effect of high-explosive anti-tank munitions. And then there is steel-encased depleted uranium that has been added for the tank's protection against high-velocity, kinetic energy penetrating munitions. The next step may involve wildly expensive compressed fiberglass, depleted uranium mesh, electro-magnetic armor, or some combination of all three.[21]

Since one of the modern main battle tank's prime roles is to destroy other tanks, these improvements in armor protection have provoked tank gun designers to upgrade the tank's main weapon to achieve greater power at longer engagement ranges. In a rather short period of time the standard American tank gun, for example, has gone from 90 mm to 105 mm, and then to 120 mm. The new Soviet 125-mm tank gun will probably force the Americans to the next step—a 140-mm gun or even an electrothermal gun.[22]

Each of these innovations in tank design has carried costs and penalties. Heavier armor and guns have demanded more powerful engines, larger transmissions, and complex suspension systems. All of this has driven the cost of the tank up, and the added complexity has increased manufacturing time. It has also greatly added to fuel consumption and the attendant problems for battlefield logisticians. Tank units must now be followed by legions of highly trained mechanics and millions of spare parts, and they must be supported by a constantly expanding training system.

The anti-tank weapons designer, in comparison to the tank designer, has a somewhat better time of it. He can afford to put much more into each generation of his relatively inexpensive systems, as escalating tank prices make it cost effective to enhance substantially each new anti-tank system. He also deals with technologies that are inherently developing faster. Sensors are as yet in their infancy, each new one marking a considerable improvement over the older varieties. Guidance systems are making great leaps in performance, providing protection from interference, achieving vastly extended ranges with fiber optic filaments, and bringing about top attack flight profiles so that the large and relatively thin-skinned upper surfaces of the tank can be penetrated. Anti-tank warheads are also being rapidly improved. It is expected that the mid-1990s will see broad use of the tandem charge, an initial penetration made by one explosive or penetrator and the succeeding penetration made by others. Lasers can be used to blind any tank crewman who is using vision blocks to see out of the tank, and the laser can "craze" vision blocks, destroying their transparency.

Additionally, there is much improvement that can be made in some of the older anti-tank weapons for minimal cost. For example, the Swiss have produced a precision warhead for the 20-year-old American Dragon missile, achieving twice the penetration of the original version. The reconfigured Dragon obtains a 1500 meter range and is guided by a day- or night-stabilized tracking system. This version, the Dragon III, is expected to be effective against reactive armor.[23] All of this is taking place while the tank designer is largely reduced to the plodding methods of adding more armor, bigger guns, and—always—more expense.

The American effort to produce the Advanced Anti-Tank Weapons System medium will probably be successful before the year 2000. The system will weigh less than 45 pounds, be capable of destroying known and projected Soviet armor, and probably use tandem warheads. The guidance system will probably take advantage of a technological breakthrough, infrared focal plane array technology. By reducing focal plane seekers to only five inches in diameter, the new hand-held system should be able to strike targets at 2000 meters with superior accuracy and reliability, day or night, and in almost all types of air obscuration conditions. It may be capable of sensing its prey through multiple means, and will not require the firer to be exposed during the flight of the missile. It may have several selectable flight paths, to include rear attack, and it will undoubtedly be cost effective, since each tank kill represents vast sums.[24]

The technical design problems associated with building man-portable missiles to destroy airplanes are actually smaller than those surrounding the development of the future anti-tank missile. Combat aircraft are designed for speed and that, in turn, means heat, which creates an easily identifiable signature in the air. These aircraft must be highly maneuverable, so that armor must be shunned or limited. Unlike the tank, the aircraft does not have any place to hide unless it resorts to standoff weapons, and then it becomes a mere missile bus.

Despite its success in the hands of Afghan insurgents against first-line Soviet aircraft, the man-portable U.S. Stinger anti-aircraft missile will probably be replaced in the mid-1990s. The Stinger POST (passive optical seeker technique) will be phased out in favor of the Stinger RMP (reprogrammable microprocessor). The new version will have the capability to identify and ignore a number of aircraft deception techniques that are frantically being added to almost all modern combat aircraft. It will use multiple seekers, improved infrared as well as ultraviolet sensitive devices. At about the same time, the man-portable version of the extremely fast British Starstreak anti-aircraft missile will be due for production.[25]

Less is known about the new Soviet shoulder-fired anti-aircraft system. The SA-16, the replacement for the older SA-7, is thought to be capable of all-aspect engagement and to have an improved warhead. It may have built-in anti-deception guards, a more sensitive sensor, and probably digital electronics. It is unlikely that the SA-16 is very far behind the improved Stinger. Both missiles can be expected to be effective and constantly upgraded.

Along with the new Stinger and SA-16, the American FOG-M (fiber optic guided missile) will likely be fielded in the 1990s. Although not man-portable, the 90-pound missile has a small TV camera (day) and infrared viewer in the nose and has a 16-kilometer range. Designed for use against both helicopters and tanks, the FOG-M provides great safety for the firer and great promise for a similar, but smaller, shoulder-fired version in the early 21st Century.[26]

A major problem with both combat aircraft and tanks is that they seem to be rapidly reaching the inevitable stage of technical development where great expense is required to achieve a marginal improvement. Another major problem is that both weapons systems are becoming increasingly vulnerable to the foot soldier's new weapons. Since the foot soldier is the prime feature of both special operations and low-intensity conflict, the infantryman's increasing relative status on the battlefield will probably bring about the increased status of both special operations and low-intensity conflict. It was the near invulnerability of both the tank and the airplane to the foot soldier's weapons that ushered in the age of mechanized warfare at Cambrai. That era may be rapidly disappearing.

The Changing State of Satellites and Airborne Collectors

Increasing vulnerability applies not only to tanks and combat aircraft, but also to the very highest forms of technical intelligence collection. It is one of the great ironies of the late 20th Century that, despite the immense sums spent on satellite and high-altitude airborne intelligence collection systems, war leaders of the world's most advanced nations will probably know less about their equally advanced enemies than their World War II, or even World War I, predecessors. After all, air photo reconnaissance by all belligerents was becoming normal in 1917, and it was certainly routine all during World War II. In an early 21st Century war between advanced nations, satellites, not cities, will probably be the first casualties. The complex post–World War II technical intelligence collection systems perform well during peace and contribute to deterrence, but once the shooting starts, other means will have to be used.

It is not that the penetration aircraft is now or soon will be prohibited from overflying enemy territory. The point is that penetration flights against a sophisticated adversary are progressively becoming more hazardous propositions, especially in a heavily defended zone. For example, American air operations over the Red River Delta during the Vietnam War were difficult enough, but that was against a comparatively crude air defense system. Consider the problem twenty years hence. While most penetration aircraft have managed only slight technological gains, air defense systems have been vastly improved. The penetrating airman is being driven to extremely expensive low-observable aircraft such as the B-2 and F-117.

The U.S. Patriot is typical of the anti-aircraft systems available to First World nations, and it is a system that is constantly being improved. The Patriot phased-array radar system can identify and track one hundred aircraft, and the system can then engage nine penetrators simultaneously with nine separate missiles. The homing missile has a slant range of 70 kilometers and can hit a target 24 kilometers above the earth's surface. This type of system will be operating in several countries during the mid-1990s and will be found in a somewhat different form within the navies of a number of nations. The extended slant range is an important feature, since this gives the air defender the capability to shoot down enemy aircraft while they are still over their own territory. The improved U.S.-German version will be autonomous—that is, after launch it will be capable of homing in on its prey without instructions from the ground radar.[27]

The attacking combat aircraft is thus driven to a high-cost, low-observable design or to a low-level penetration flight profile. However, lower altitudes are also becoming increasingly hazardous. Both the Soviet Union and the United States have developed highly capable Airborne Warning and Control Systems (AWACS) that fly above and locate low-flying aircraft, passing off targets to interceptors that have a look-down, shoot-down capability. In the space of twenty years, the air defender has gained far more effectiveness than has the manned air attacker, and that trend can be expected to continue.

For these reasons, future penetration flights against an advanced air defense system will be a very restricted activity. For example, traditional photo reconnaissance missions will not be likely. A target worth the risk of photographing will probably be one that is so heavily defended that it would be inefficient to send a manned aircraft simply to gain knowledge. If such a target were to be overflown, it would have to be reached through a very carefully planned strike, complete with a considerable amount of deception, numerous decoys, and a heavy air defense suppression component. Armed reconnaissance will probably be out of the question. Overflights of very sensitive installations or field armies may well be in the impossible category.

However, penetration flights against an advanced nation's air defense system will probably be possible in less defended regions. Despite the great advances in these defense systems, they do have limited ranges and are so expensive that they are sparingly deployed, providing protection only for a nation's critical forces and most important cities and facilities.

The rising vulnerability of manned aircraft is matched by a similar trend with satellites. Within the decade of the 1990s, it is expected that the Soviet Union will develop ion sources and radio frequency quadrapole accelerators to the point that particle beam weapons will be capable of disrupting the operation of most intelligence collection satellites.[28] And, particle beam weapons are only one way to attack satellites. Space-based collectors are also vulnerable to physical disruption by an interceptor satellite or one using a laser gun. They are also susceptible to electromagnetic jamming of their essential communications links, and they can be destroyed or damaged by the simple placement of debris in their paths.

The great utility of satellites is in their peacetime intelligence verification, communications, and initial warning roles. The emergence of radar-imaging satellites in the late 1980s has considerably enhanced the value of space-based intelligence platforms, essentially nullifying the effects of adverse weather. By the late 1980s, eight nations had developed satellite manufacture and launch capabilities, but many more nations were using their own space platforms, because this activity had been transformed into an international commercial service. By the turn of the century, seventeen countries will likely be maintaining twenty-three sophisticated intelligence collection photo satellites with a resolution quality of one meter.[29] However, the nation that depends on satellites for obtaining vital information or communicating during the conduct of a high- or mid-intensity conflict is a nation that is banking on a special and possibly improbable condition. Like manned airborne collectors, satellites are vulnerable.

National leaders are not likely to long tolerate the prospect of being uninformed during a mid- or high-intensity conflict, as they surely will be if they rely on airborne and satellite intelligence systems. Prudent leaders will invest in a ground reconnaissance capability, as some have already done with the formation or enhancement of special operations forces. These units have growing effectiveness and are not plagued by increasing vulnerabilities.

Technology, Insurgents, Commandos, and Terrorists

On balance, technological trends are favoring the terrorist, the special operations unit, and the insurgent as the 21st Century approaches. This is particularly true in the realm of state-supported terrorism and the externally assisted insurgent. In some respects, change is the result of the increased effectiveness of new items that a human can carry or wear—shoulder-fired weapons, rations, communications equipment, devices for assassination and mass terror, and body armor. It is, perhaps, a logical and predictable technological reaction to the age of mechanized warfare. That era began with man using machines to transport himself and his weapons into battle. It developed into a contest between manned machines and is moving into an age where man dismounts his machines and sends them into battle. In part, man must divorce himself from the machine because it is increasingly vulnerable to his opponent's weapons. But, man's safety is not the only consideration.

Competitive manned war machines are rapidly becoming prohibitively expensive. There are steadily increasing costs of sophisticated manufacturing facilities, training time, complex supply systems, maintenance, and the ever present necessity to support a continuous technological development process that hopefully counters obsolescence. By the late 20th Century, the cost of first-line manned war machines has become so great that the list of nations that can afford them is diminishing.

There are, therefore, many reasons and many trends that are pointing to a new era of warfare. The technologically oriented observer readily points to more expense and more sophistication: the new age will be one of robot warfare. But, there are others who indicate a somewhat less complicated but just as revolutionary new age, an age where men who stand on the ground with weapons in their hands defeat mechanized armies. This view has obvious implications for the future of low-intensity conflict and even greater implications for the higher end of the conflict spectrum and the use of special operations forces.

Notes

[1] Paul F. Gorman, *Supporting U.S. Strategy for Third World Conflict* (Washington, D.C.: Regional Conflict Working Group, U.S. Department of Defense, 1988), p. 16.

[2] Chris Bellamy, *The Future of Land Warfare* (New York: St. Martin's Press, 1987), pp. 181–184.

[3] Colonel John T. Burke, "The Firepower Revolution," *Army* (November 1988): 50–60.

[4] Major Roy A. Griggs, "Maritime Strategy on N.A.T.O.'s Central Front," *Military Review* (April 1988): 54–65.

[5] Submunition results as reported in *Naval Institute Proceedings* (August 1988): 124. Tomahawk enhancements were reported to the U.S. Congress by Rear Admiral William Bower, as quoted in *Jane's Defence Weekly* (May 7, 1988): 881.

[6] Commander-in-Chief, U.S. Special Operations Command General James J. Lindsay, "The C^3I Challenges for U.S. Special Operations Command," *Signal* (May 1988): 21–23.

[7] Roger Kohn, International Maritime Organization, as quoted in an Associated Press news release from London, November 11, 1988.

[8] Lieutenant Colonel Paul L. Blackburn, U.S. Army War College paper, "Weather Modification as a Weapon of War," Carlisle Barracks, Pennsylvania, 1975, pp. 3–5.

[9] U.S. Department of Agriculture, Forest Service, *Forest Fire as a Military Weapon* (Washington, D.C., 1970), p. 73.

[10] U.S. Chief of Naval Operations Admiral Carlisle Trost, as quoted in *The New York Times,* July 31, 1988, p. 21.

[11] Commander-in-Chief, U.S. Special Operations Command General James J. Lindsay, as quoted by George C. Wilson, "Navy Acquires Minisubs for Secret Operations," *Washington Post,* March 3, 1988, p. A8.

[12] Group Captain Peter Millar, "The Central Region Layer Cake: An Essential Ingredient," in Major General J. J. G. Mackenzi and Brian Holden Ried, eds., *The British Army and The Operational Level of War* (London: Tri-Service Press, 1989), p. 27.

[13] Tobin Carter, "Daisy Cutters," *Vietnam* (Summer 1988): 12–14.

[14] J. Wallace Higgins, "Porterage Requirements and Tables" (Santa Monica, CA: The Rand Corporation, 1967), pp. 37–48.

[15] Philip Brandler, Acting Director, Food Engineering Directorate, U.S. Army Troop Support Command, as quoted in *Army Times* (April 10, 1989): 30.

[16] Captain Peter E. Blaber and Chief Warrant Officer Kevin M. Egan, "MPBS: The Multi-Purpose Boot System," *Infantry* (March-April 1989): 14–17.

[17] Walter B. Wriston, "Technology and Sovereignty," *Foreign Affairs* (Winter 1988/1989): 63–75.

[18] Author's interview with Dr. Robert F. Bornstein, September 16, 1988. Also see his paper "Potential for Use of Subliminal Techniques as Propaganda Tools" (Gettysburg College, PA, 1988), pp. 26–27.

[19] Steven M. Shaker and Alan R. Wise, *War Without Men: Robots on the Future Battlefield* (McLean, VA: Pergamon-Brassey's, 1988), p. 98.

[20] Major General Perry M. Smith, "Air Battle 2000 in the N.A.T.O. Alliance: Exploiting Conceptual Technological Advances," *Air Power Journal* (Winter 1988): 4–15.

[21] William Haskell, U.S. Army Materials Technology Laboratory, as quoted in *The Wall Street Journal,* January 9, 1989, p. B4. Lieutenant Colonel John Reitz, as quoted in *Army Times,* April 10, 1989, p. 28.

[22] General Donn A. Starry (U.S. Army, Ret.), as quoted in *Armed Forces Journal International* (February 1989): 68–76. Also see *Defense News* (May 8, 1989): 1.

[23] "U.S. Marines to Upgrade Dragon," *Jane's Defence Weekly* (November 26, 1988), p. 1350.

[24] Colonel Earl Finley, Project Manager, AAWS-M, as quoted in *Army Times,* November 28, 1988, p. 29. Glenn W. Goodman, Jr., "New Army IR-Guided Missile to Rely on Focal Plane Arrays in the 1990s," *Armed Forces Journal International* (May 1989): 72.

[25] David G. Harris, Army Missile Command, as quoted in the *New York Times,* March 27, 1989, p. 15. "Army Looks into Buying a Companion for the Stinger," *Army Times* (April 10, 1989): A24.

[26] Colonel Oleh Koropey, FOG-M Program Manager, as quoted in *Army Times* (December 26, 1988): 26. Also see the comments of Thomas Jarrell, Boeing Company Program Manager, FOG-M, in *Jane's Defence Weekly* (December 10, 1988): 1447.

[27] Caleb Baker, "Patriot's Accuracy, Range to Be Increased," *Army Times* (April 24, 1989): 24.

[28] U.S. Department of Defense, *Soviet Military Power* (Washington, D.C.: USGPO, 1986), p. 47.

[29] Owen Thomas, "Nations Keep an Eye on Each Other," *Christian Science Monitor,* September 28, 1988, p. 3.

CHAPTER FOUR

Special Operations in High-Intensity Conflict

The specter of a final battle between East and West resulting in the destruction of civilization diminishes with each passing day. The literature on the conduct of nuclear war is becoming less plentiful, and there is little remaining popular interest in civil defense, fallout shelters, and survival stocks. These concerns have been replaced by speculation on exchanges of data, arms control negotiations, verification procedures, and mutual destruction of nuclear delivery systems.

Yet the number of nuclear-capable nations is likely to grow and, with proliferation, nuclear war may become more probable. The reason for the lack of attention devoted to this danger is that, in the realm of nuclear war discussion, there is an almost single-minded fixation on the nuclear inventories of the two 20th Century superpowers. There is a following common belief that any nuclear incident between the United States and the Soviet Union would result in the wholesale launching of every nuclear weapon in the arsenals of these two nations. This appalling thought overshadows all else, and too little notice is taken of lesser and far more likely scenarios. Both Israel and Iraq will, in all likelihood, be capable of delivering nuclear strikes on one another in the mid- or late 1990s, so too India and Pakistan. How would these conflicts be conducted? There is little consideration given the subject. There is also scant concern over the vital status of northern hemisphere war deterrence once Moscow and Washington agree on sizable reductions in the numbers of intercontinental ballistic missiles.

Averting Armageddon in the 21st Century

The lower limit of the mutual reductions in Soviet and American strategic nuclear forces will probably be determined in Beijing. The Soviet Union is unlikely to decrease its numbers of extended-range weapons to a point of near equality with those of China. But, should Soviet-U.S. reductions appear as an unqualified success, strategic arms limitation talks between Moscow and Beijing are probable. The next lower limit may be determined by the combined French and British nuclear arsenals. Moscow is not likely to allow the West European nuclear capability to overawe its own arsenal. Strategic nuclear weapons will undoubtedly exist in the late 1990s, but there will probably be a rough balance among the major nuclear powers. The resulting problem is that bi-polar nuclear terror may be less of a means to maintain the peace in the northern hemisphere.

There are at least three natural products of a reduced reliance on strategic nuclear weapons. The first makes conventional forces more important in the N.A.T.O. and Warsaw Pact equation. The troublesome aspect of this situation is that it raises the prospect of an aggressor actually conquering something worth having. A massive employment of strategic nuclear weapons at least has the saving grace of total, mutual destruction and the absolute lack of any reasonable reward for an initiator of violence. In short, strategic arms reductions have the happy effect of diminishing the unthinkable, but at the expense of raising limited war to the distinctly thinkable.

A second condition is an increase in the degree of caution that the Soviet Union and the United States display toward one another during periods of crisis. Both states have steadily improved their records of avoiding "all or nothing" situations, and there will be all the more reason to sidestep confrontations in an age when an economically strong, nuclear-capable West Europe confederation is pursuing its own interests. It will be unlikely that an American president will again risk the devastation of North America by assuming the role of protector of Western Europe. The U.S. chief of state will have the policy option of allowing Western Europe and the Eastern bloc to sort out their own disputes. Rational leaders in both Washington and Moscow are apt to engage in fewer gambles.

All of this brings about a third condition: a Western search for non-nuclear means of deterrence. Warsaw Pact nations are likely to rely on and maintain their decided edge in conventional forces. President Mikhail S. Gorbachev's 1988 unilateral conventional force reduction announcement, for example, was calculated to leave pact forces with a 20,000 tank and 21,000 artillery piece superiority over N.A.T.O.[1] And, the announcement was made in the midst of a massive Soviet conventional arms modernization program. Between the time Gorbachev assumed power and the time he made his announcement, the Soviet leader supervised a massive modernization program that added more new tanks and artillery pieces to Soviet forces than existed in the combined armies of France, Britain, and the Federal Republic of Germany.[2] Reduced reliance on strategic nuclear arms to maintain deterrence is not likely to be compensated for by parity in conventional forces. Successive N.A.T.O. commanders have tried and failed, over a period of thirty years, to achieve a conventional force balance in Europe.

While conventional forces do not offer the West a realistic and adequate deterrent, unconventional forces may. The West has an increasingly potent, but as yet only potential, arrow in its quiver: the capability to exploit dissidence within the western and southern republics of the Soviet Union and in East European nations. Moscow's power is fundamentally based on its ability to control these peoples, and any threat to that control could cripple the Soviet Union. Moscow's leaders are particularly alert to this possibility.

Gorbachev announced his approach to the problem in 1987. Recognizing there was "a certain section of the people" who might "descend to nationalism," the Soviet leader stated that there were people in the West who might try to "undermine the friendship and cohesion of the Soviet Union." He then proclaimed that "Soviet law stands on guard, protecting the accomplishments of Leninist national policy."[3] A year later, the head of the KGB was given the task of rewriting Soviet laws. This official, Victor M. Chebrikov, described KGB duties as including the functions of preventing "crimes against the state" and "helping the deviant to shed his delusions." Less than a year later, the new laws began to emerge. The Supreme Soviet passed several tough measures to deal with the country's growing irredentism. For example, those Soviet citizens who make "public insults against the Soviet Union" face prison terms ranging from three to ten years.[4] Additionally, a part of Gorbachev's "restructuring" program revised the Kremlin's internal security apparatus. Chebrikov the "lawgiver" was promoted to the rank of central committee secretary and given oversight responsibilities for the KGB, the Interior Ministry, and the Justice Ministry. The new KGB head, Vladimir Kryuchkov, was perhaps chosen for his expertise in dealing with dissent against Moscow by the peoples of the Warsaw Pact. Kryuchkov had been a KGB agent in Budapest during the 1956 Hungarian uprising and had attended the infamous "dinner party" where the "guests," rebellious Hungarian leaders, were arrested and some of them were later executed.[5] Soviet leaders clearly recognize their nation's vulnerability. However, turning unrest and dissidence into a Western non-nuclear war deterrent requires the West to debate the subject, devise policies, and substantially enhance its special operations forces.

Since the mid-1950s, the West has maintained an uncoordinated and essentially passive position on civil unrest within the nations of Eastern Europe and inside the Soviet Union. Despite considerable evidence in open print, there has been official silence on American support of dissident groups in the early 1950s and both the Eisenhower and Johnson administrations found themselves on the defensive during and after the revolts in Hungary and Czechoslovakia. What should have been a strength was made into a Western weakness. During the 1980s, there was some change in the actions of Western governments, notably in regard to Jewish emigration from the Soviet Union and in matters relating to the Polish Solidarity movement. However, official Western consultations over East European and Soviet unrest have been episodic and indeterminate. Presented time and again with an undeniable Soviet vulnerability, the West most often stood helpless behind its nuclear shield. In periods of East European unrest, tension has increased, but it has been Moscow and not the West that held the initiatives. The West has simply accepted what the Soviets caused.

The absence of Western debate and consensus on Soviet irredentism and East European separatism has played into the hands of those who would oppose any Western initiatives in this field. Some argue that rapid, uncontrolled destabilization of the Marxist empire holds danger for the West.[6] Their reasoning usually follows a line that holds the Soviet empire is one built on a "foreign devil" theory and that Moscow's leaders could be expected to take irrational and possibly drastic action if serious political fragmentation occurred. There is some historical support for this view, the idea that an East-West war could result from a destabilized Eastern bloc. One study of the origins of modern war concludes that wars have sprung more from apprehension than from attempts at conquest. Another study concludes that only appeals to emotion or impulses can support the essential spirit for war. Logic and reason are not enough.[7] A destabilized Soviet bloc is likely to cause Moscow a considerable amount of emotion and apprehension, and perhaps a deadly impulse.

However, that argument ignores the fact that the West is surely going to be blamed in any event. A U.S. analysis of Moscow's reaction to internal unrest concludes there is a consistent Soviet claim of "imperialist meddling."[8] This reaction persists with those new leaders who are attempting to "restructure" Soviet society.[9] The absence of a clearly understood, publicly stated Western policy in this area makes it almost certain that internal revolt within the Soviet empire will result in an active anti-Western action program by Moscow. Not only have Soviet leaders maintained the initiative in this arena, there is every indication they will continue to do so in the future.

A Western policy on the Soviet empire's internal fragmentation is needed. The policy might state that under the conditions of peace in Europe, violent political change, both within the Soviet Union and among its East European allies, represents a threat to the West, and only peaceful change can be welcomed. The policy's corollary could state that support and direction of internal revolt within the Soviet empire during a European war is absolutely essential to the survival of the West.

Some in the West have suggested that internal bloc unrest should be exploited in any case, under any conditions, during war or peace.[10] Few proposals could be more risky. The capability to support insurgency in the Soviet Union and bloc nations should be squarely placed in the hands of Western military authorities as a publicly recognized and vital component of Western military deterrence. It should not be used as a tool in a dangerous game of peacetime manipulation of Moscow's internal tensions.

In order to integrate fully insurgency support into the arsenal of Western deterrence and enhance its value, another new policy would be useful. There should be a flat Western announcement that, in the event of war, no population centers outside of the Russian Federated Socialist Republic are to be targeted by Western mass destruction weapons. The simple and following conclusion is that East-West problems hinge on the actions of one ethnic group—the Great Russians. Ukrainians, Byelorussians, citizens of the Moslem republics of the U.S.S.R., and other non-Great Russians would be regarded by the West as potential wartime neutrals or even allies, peoples that must be treated differently. This proposed policy recognizes a well-understood facet of Soviet life, that leadership positions within the party and Soviet government are largely held by Great Russians. Such a policy would isolate and define the essence of the major problem between East and West.

Whether or not new policies are adopted, the West should (at least) carefully examine its own capabilities to exploit the internal weaknesses of the Soviet empire. The chief deficiencies in the Western structure to support and direct the energies of anti-Soviet elements within the U.S.S.R. and Warsaw Pact nations are: an inadequate force structure, lack of enough language-capable Special Forces soldiers, an inadequate air penetration special operations support element, and the absence of a coordinated special operations command and control arrangement. The strengths of the Western insurgency support structure are: an existing military organizational base, a school system for language and skills training, available doctrine, and a materiel research and development structure. Unfortunately, these strengths are almost wholly American; European nations contribute little to the overall Western capability. Even considering the deficiencies, Western potential in this field grows as Soviet irredentism and East European separatism increase. Yet, that potential can be substantially increased and can become a powerful element of conflict deterrence.

American military capabilities to support and direct insurgency are largely contingent on the readiness and availability of its Special Forces groups, one of these active-duty organizations being assigned to each of the regional U.S. unified commands. The group designated for the European theater, and the one that would be immediately involved in combat operations in the event of war there, is identical to the other Special Forces groups and is built around its fifty-four "A" detachments,

12-man units that are to be employed behind enemy combat forces.[11] During the Vietnam War, the "A" detachments normally supported and directed the activities of about 300 indigenous fighters. It is doubtful that they would be capable of handling many more in a European guerrilla support role. Assuming a conservative 25 percent attrition rate upon initial infiltration into Eastern Europe at the outset of war, the Special Forces group would be capable of supporting and directing the efforts of approximately 12,000 insurgents.

The only other Western military organization that might be tasked with an insurgency support role is Britain's 22nd Special Air Service Regiment. A relatively small unit, the regiment may add another 8,000 or so to guerrilla ranks operating behind Warsaw Pact forces. That, of course, assumes that the SAS could find the necessary transport to reach appropriate operational areas. All told, the Western military ground force structure could probably support little more than 20,000 insurgents in the initial phases of an East-West conflict in Europe. That figure, 20,000, is undoubtedly less than the normal anti-Soviet resistance potential in Poland alone and is not likely to pose serious concerns in Moscow.

Despite the modest nature of the Western ground insurgency support capability, it is more than can be supported by the companion special operations air element. An optimistic estimate of the available air penetration transports in the mid-1990s totals about 70 airframes that have or will have the necessary range, navigation, and electronic deception equipment. These aircraft would have to support both American and British forces, since the Royal Air Force has no special operations units. A rough calculation of team infiltration and resupply sorties, minus the inevitable attrition figures, produces a shortfall of about ten aircraft.

The current and projected inventory of special operations aircraft destined to transit pact airspace is not only insufficient in numbers, it does not contain the technological diversity required to sustain itself in combat. The growing Warsaw Pact air defense system will surely take a heavy toll of American special operations penetration aircraft in the future. However, technology exists to complicate greatly the efforts of pact air defenders and to enhance the capabilities of N.A.T.O.'s insurgency support forces.

The unveiling of the American B-2 and F-117 stealth, or low-observable, aircraft in the late 1980s produced another dimension and another set of problems for the Warsaw Pact air defense system of the 1990s and early 21st Century. Considering the U.S. combat air fleet alone, penetration aircraft expected to operate behind front-line Warsaw Pact forces during that time frame include the B-52, B-1, B-2, F-16, F-111, F-15, F-117, MC-130E, MH-53J, and an array of ground-hugging cruise missiles. Each of these penetrators has a different mission flight profile, electronic signature, electronic and other deception means, and particular area of the battlefield where it is expected to be found. All of this requires Soviet planners to invest in a bewildering number of sensors, interceptors, and missiles—an extremely complex, costly, and ever-increasing air defense system.[12] These facts add up to a considerable Western war deterrent and for that reason, if for no other, Soviet air defense planners should be presented with yet another problem.

As far back as 1973, the U.S. Air Force had the Windecker YE-5A delivered; it is a single-engine, propeller-driven airplane that is extremely difficult to detect by radar. The small craft was constructed with a dielectric fiberglass outer "skin" and used radar-absorbing materials for interior strength members. Since that time, a wide variety of radar "transparent" materials such as Kevlar, carbon fiber, and the ultrastrong and heat-resistant carbon-carbon have been made available to designers. Additionally, great progress has been made in the field of ceramics, so that reliable power plants that have almost no radar returns are quite possible. Then too, radar absorbent materials have been made, materials that can blanket those parts of an aircraft or its contents that must be made of metal.[13] There is every reason to believe that a small, special operations air transport could be easily constructed of the latest low-observable materials.

Aside from enhancing the chances for survival of Western special operations forces, a special operations low-observable transport has a number of attractions. Having its own penetration profile and characteristics, it would further complicate the Warsaw Pact air defense problem. Also, this type of aircraft need not be as expensive as the F-117 and B-2, whose development programs bore the initial research and engineering costs for the technology. The transport could be relatively small, involving only twelve passengers and about 500 pounds of cargo. It would also not have to have any high-performance characteristics associated with the earlier two "stealth" aircraft and could be constructed of very light, radar transparent and radar absorbent materials. Wood and fabric should not be out of the realm of consideration.

Low-observable penetration aircraft are no panacea, and this field of military technology must take into account detection by infrared sensors and a number of other identification devices. For example, stealth aircraft may be detectable by specialized radar, albeit at considerable expense on the part of the defender. Some experts claim an over-the-horizon, backscatter radar system will detect the top-positioned engine intakes of a low observable aircraft in a 1,500-kilometer-wide zone, 1,000 kilometers forward of the transmitter. Careful placement of the backscatter transmitters could provide coverage of border areas.[14] But, once again, this type of detection system is yet another high expense for the defender.

Although the current and planned Western ground and air capability to support insurgency within the Warsaw Pact nations does not appear to be sufficient to have a substantial war deterrent value, a five-fold increase probably would. Moscow would then have to plan on the possibility of having 100,000 armed guerrillas in the rear of its forces in the beginning stages of a European war. The increased cost to the West would amount to about 4,000 additional special operations troops and 330 new transport aircraft, planes that should have a reasonable chance of penetrating pact air defenses.

Another necessary enhancement in the Western insurgency support structure is a much higher degree of language proficiency among those soldiers designated to work behind enemy lines. The need here has little to do with the esoteric field of agent preparation. Western special operations forces cannot be expected to risk exposure to bloc security services in areas where resistance elements are not strong. The fundamental requirement is simply to acquire enough working vocabulary to pass on weapons instruction and operational techniques, and sufficient language comprehension to receive and understand local intelligence reports. Despite these minimal needs, few active-duty and practically none of the reserve component special operations elements, British or American, have basic foreign language qualifications. Traditionally, manual skills training, the easier task when compared to the year-long schooling required for most East European languages, has taken precedence over language training. A simple reversal in priorities would suffice to rectify this deficiency. Language qualification must be made a prerequisite to receiving skills training.

Another step necessary to enhance Western insurgency support capabilities centers on the thorny matter of command and control. N.A.T.O.'s strategy is essentially defensive, and some member nations have avoided the expense of creating and maintaining an offensive element destined to be used in the enemy rear.[15] The end result has been that only the United Kingdom and the United States have any actual capability in the field, and both are understandably reluctant to subject their plans and programs to the scrutiny of allies who presumably have little expertise in such matters. Therefore, the West not only has no coordinated insurgency support effort, it has no combined command and control mechanism for the activity. Each Western state is thus on its own in the matter of insurgency support, and what potential exists is almost wholly American. If the American role in the forthcoming West European confederation is diminished, which it probably will be, the West may well lose an important element of war deterrence. Coordinated Western policies, enhanced Western special operations forces, and a combined command and control arrangement are clearly needed.

Although the value of this deterrence option, insurgency support, may be a bit difficult to calculate, it is no more ambiguous than the value of the strategic nuclear deterrent itself. In order for an insurgency support capability to constitute a worthy element of deterrence, Moscow must first be convinced that some of its people in the western and southern republics and some of its allies' citizens would take up arms against the Kremlin. Soviet leaders would then have to believe that the West could infiltrate insurgency support teams and supplies into areas of public discontent. Given those beliefs, a Western insurgency support force would have some deterrent value.

On the other hand, the value of the strategic nuclear deterrent is dependent on Moscow's perception of the West's ability to reach a timely, irretrievable, and awesome determination to launch nuclear-tipped missiles, transmit the necessary commands, and have those commands obeyed. Additionally, the Soviets must be convinced that the reliability of Western missiles and reentry vehicles is sufficient to guarantee ignition, launch, and accurate navigation, and that the weapons will detonate on reaching their targets. If each of these six steps had a 90 percent probability of occurrence, the Soviets could expect that less than 50 percent of the Western-selected targets would be destroyed. The Soviet Union has been heavily damaged before. There is a base of experience. In the first blush of Hitler's 1941 invasion, Moscow lost control of about half of its population and two-thirds of its industry. Soviet leaders, well schooled in the history of their "Great Patriotic War," are not apt to forget that their nation not only survived that seemingly devastating blow, but ultimately triumphed.

In essence, creating deterrence is the act of instilling fear in one's adversary, and fear is basically perception. No one in the West can determine with any degree of accuracy what the Soviet leadership at any given time is likely to fear most. Moscow's fear of armed, internal revolt may be as high or higher than its fear of nuclear war. And this important Western war deterrence option, insurgency support, cannot be matched by a similar Soviet capability. The absence of a hostile citizenry is a natural strength of democracies. In an era of "competitive strategies," where security planners seek to forward programs that build on strengths and exploit vulnerabilities, the West is giving short shrift to a powerful strategic option.

Spetsnaz in High-Intensity Conflict

If the Soviet Union is involved in a nuclear war during the 1990s or early in the 21st Century, its probable adversary will be the West, possibly China, and conceivably both. In any case, its Spetsnaz forces will be used, but the rapidly diminishing appeal of Marxism will ensure that these special operations elements will not be employed in an insurgency support role. Soviet Spetsnaz, army, and navy will be used deep in the rear of Moscow's enemies as commando and reconnaissance units.

Likely missions for Spetsnaz include raids on headquarters and naval facilities. Some authorities include assassination on the list of Spetsnaz tasks and state that these forces will appear in the uniforms of their enemies.[16] In a nuclear conflict of massive proportions, this type of threat is probably of least concern to anyone; however, in a limited nuclear war, Spetsnaz capabilities must be taken seriously because there are indications that these forces have a good chance to reach their destinations and may have already visited their assigned targets.

Spokesmen for Stockholm's International Peace Institute claim some 20,000 Soviet special purpose troops visit the West each year, many traveling aboard the 100,000 Eastern bloc trucks that enter the Federal Republic of Germany annually. Swedish government officials report Spetsnaz troops often pose as students selling art.[17] Although there is no identified Soviet special operations air element, one may exist, and it would possibly use a rather antiquated but effective 1947 vintage biplane to deliver parachutists in wartime. The AN-2 is a single-engine troop carrier with fabric-covered wings and tail surfaces, giving it an extremely small radar cross-section. This aircraft is capable of speeds as low as 40 miles per hour, speeds that may not register on some gated, Doppler-type radar systems. Configured to carry fourteen paratroopers, the AN-2 can tow several gliders and can be rigged with floats for water landings. Although only about 290 of these aircraft are counted by Western Soviet watchers in bloc military inventories, well over 1,000 have been produced; many of them are within Moscow's "civil" airline, Aeroflot. The existence of the AN-2 Colt is of serious concern to some U.S. defense officials.[18]

Despite the apparent effectiveness of Spetsnaz in some selected high-intensity conflict scenarios, Soviet capabilities in behind the lines operations do not compare favorably with similar Western capabilities. On this end of the conflict spectrum, the premium is on deterrence and, without a realistic hope of managing the efforts of insurgents in Western Europe, the U.S.S.R. has to be satisfied with raids and reconnaissance, operations of minimal deterrent value. And, if the West fields an enhanced and coordinated insurgent support force, Spetsnaz units may have to drop their offensive plans and be relegated to a rear area counterguerrilla role. The West has the inherent advantage here, and it is not the only one within the realm of special operations.

Civil Affairs and Psychological Operations

The strength and potential of a military or naval force are largely dependent on the talent and resources placed at its disposal. Most of the world's armed forces contain less than 2 percent of their nation's population. An old and trusted rule of thumb holds that 10 percent of any nation's population is about the maximum strength a military force should reach before its country's industrial, agricultural, and financial fortunes go into ruinous decline. However, it is a matter of record that about 50 percent of a nation's gross national product can be devoted to the support of a military force for periods of four to five years without irreversible negative effects. It is therefore in a nation's military interest to expand its GNP, and that means it must devote the minimum essential amount of its manpower and economy to its military forces in peacetime.

The potential military power of any nation is therefore contingent on the speed and efficiency with which it can expand the ranks of its forces and marshal resources for their disposal. Security planners have the unenviable task of calculating all of this and balancing a military commitment against similar figures of a potential adversary.

Planners know that the best type of security asset is one that does double duty, contributing to GNP growth in peace, but having instant military utility in periods of crisis. These assets include food and medical service industries, communications organizations, storage and transportation industries, construction corporations with capabilities to repair and maintain transportation nets or create barriers, commercial ship repair and servicing institutions, fuel storage and transportation industries, and public organizations that can perform traffic control and police functions.

Quickly obtaining the military benefits of these civil entities is increasingly becoming the function of an American special operations component, the Civil Affairs units. Organized in World War II with the original task of military government, these formations began a functional transition in the 1980s aimed at marshaling foreign civil resources to support U.S. military forces stationed abroad. The activity, called "host nation support," is actually a familiar peacetime function that most U.S. military organizations had managed for some time. However, host nation support was normally arranged on a temporary contractual basis to take advantage of local efficiencies for the use of a particular military installation during peace. The massive orchestration and employment of foreign civil resources in support of combat operations on a theater-wide scale were not envisioned, and in any event could not be planned due to austere staffing of U.S. logistical organizations. Civil Affairs units are now beginning to provide that staffing, planning, and coordination.

Another unique facet of American special operations is the psychological operations structure. Equipped with mobile transmitters and printing presses, U.S. "psyops" units can influence public attitudes within both enemy and friendly territory by the use of radio or television broadcasts or by air-delivered leaflets. This capability will probably be of limited utility during the very early stages of a high-intensity conflict and of no use at all during a nuclear exchange. However, both the American civil affairs and psychological operations capabilities may be of the utmost importance in the post-conflict stage.

Some nuclear wars may be "winnable." The world has fortunately never experienced a war between two or more powers that have each used nuclear weapons, so experience is of no aid in envisioning this type of tragedy. Depending on the numbers of weapons used and the extent of destruction, the residual power of one belligerent may be sufficient to allow the imposition of its will on an adversary. In that case, the ability to persuade survivors and control remaining resources would be essential to consummating a victory. Those capabilities exist within the structure of U.S. special operations forces.

Substantial military psychological operations and civil affairs units are unique to the American force structure. The types of functions that these U.S. organizations perform are normally only staff responsibilities in other armed forces, those of the Soviet Union included. There is a tactical psychological operations capability in the Red Army, but it is not likely that Moscow's leadership would trust its armed forces with a potent psychological operations structure. Within the U.S.S.R. that vital mission is reserved for the Communist Party, and the prime targets are the Soviet public and the citizens of its allies. The Soviets do have an admirable ability to marshal civil resources for war use but, again, the effort is controlled by the party and its allied political counterparts. There is no reason to believe that Warsaw Pact nations can match these American capabilities on foreign soil without the existence of strong, indigenous communist parties, institutions that are disappearing.

Second and Third World High-Intensity Conflict

There does not seem to be much of a role for special operations forces in Second and Third World nuclear wars. In fact, if these events do occur in the late 1990s and early 21st Century, the results of the conflicts may be determined by outside parties. Some authorities believe that a successful use of nuclear weapons in a regional conflict could lead to an accelerated proliferation of nuclear powers. Conversely, they hold that a nuclear event causing the loss of thousands of civilians without any appreciable military effect could dramatically lower the estimates of nuclear weapons effectiveness.[19] The first alternative is one that none of the major powers would accept for a number of reasons. It is therefore likely they would take unprecedented steps to exert control over the conflict, coordinating their actions.

Coordinated intervention of the great powers into a regional high-intensity conflict could only be done, in all probability, on the basis of returning to a political *status quo ante bellum,* ensuring that no advantage accrued from a lesser power's use of nuclear weapons. The likelihood would therefore be the employment of peacemaking forces and overwhelming military restraints placed on the belligerents.

Special Operations and Nuclear War

The dominant concern in the arena of high-intensity conflict in the late 1990s and early 21st Century will probably remain squarely centered on deterrence. Strategic arms reductions will likely spawn efforts to find non-nuclear means of deterrence, because there is a point in these reductions where a limited nuclear war becomes "thinkable." Other forms of deterrence would then be needed.

The most likely area to witness a nuclear war, however, is not in the regions of North America, Europe, China, or the Soviet Union. That type of conflict is more likely to occur in the nations of the Second and Third World. These regions have not had the benefit of forty years of nuclear deterrence and near-continuous strategic arms negotiations. But, high-intensity conflict in the Second and Third World would probably bring on the intervention of the world's major powers in a cooperative effort to impose stability. It is doubtful that any nation would gain from this type of conflict. Special operations forces are of limited utility in this type of situation.

However, Western special operations forces have considerable potential in the realm of East-West nuclear war deterrence. A ready Western capability to support and exploit growing irredentism and separatism in the U.S.S.R. and within Warsaw Pact nations could become a powerful tool of war deterrence. In many ways, this form of deterrence is superior to deterrence provided by strategic nuclear weapons. A strong Western special operations capability is not apt to draw the heavy domestic criticism that Western leaders have to confront on a daily basis in defending nuclear-capable forces, their stationing, and their high costs. A Western special operations force prepared to exploit discontent within Marxist states would be a tool of deterrence dedicated to the rights of peoples who have long been oppressed. It would be a force in concert with Western ideals.

Should strategic arms reduction reach the point where nuclear war in Europe becomes survivable, American Civil Affairs and Psychological Operations forces would have moderate utility, particularly in a post-conflict stage. Taken together, U.S. special operations forces, Special Forces, Civil Affairs and Psychological Operations units, and special operations air elements have considerable utility on the higher end of the conflict spectrum. They provide capabilities their adversaries cannot match. Unfortunately, most of America's European allies have not seen fit to invest in these types of forces.

Notes

[1] Message, Headquarters, Department of the Army. Subject: General Secretary Gorbachev's Announcement at U.N. General Assembly (Washington, D.C., December 22, 1988).

[2] General John Galvin, Supreme Allied Commander, Europe, as quoted by Seymore Weiss, "A European Arms Pact?" *New York Times,* March 11, 1989, p. 27.

[3] Mikhail S. Gorbachev, *Perestroika: New Thinking for Our Country and the World* (New York: Harper and Row, 1987), pp. 119–121.

[4] Victor M. Chebrikov, Chief of the KGB, as quoted by I.F. Stone, "The KGB's Lofty Humanist," *New York Times,* October 24, 1988, p. A17. David Remnick, "16 Killed at Rally in Soviet City," *Washington Post,* April 10, 1989, p. A1.

[5] Jerrold L. Schecter, "The Soviet's New Top Cop," *Washington Post,* February 5, 1989, p. D1.

[6] George F. Keenan, "Just Another Great Power," *New York Times,* April 9, 1989, p. E25. David M. Abshire, *Preventing World War III: A Realistic Grand Strategy* (New York: Harper and Row, 1988), p. 289.

[7] A. J. P. Taylor, *How Wars Begin* (New York: Atheneum, 1979), p. 15. Quincy Wright, *A Study of War,* 2d ed. (Chicago: University of Chicago Press, 1965), pp. 289–290.

[8] Caspar W. Weinberger, *Soviet Military Power* (Washington, D.C.: U.S. Department of Defense, 1987), p. 131.

[9] Gorbachev, *Perestroika,* p. 163.

[10] Douglas Whitehouse, "Coping with Alternative Soviet Futures," in Perry M. Smith, et al., *Creating Strategic Vision: Long Range Planning for National Security* (Washington, D.C.: National Defense University Press, 1987), p. 127.

[11] U.S. Army, *Field Manual 31-20: Special Forces Operations* (Coordinating Draft) (Fort Bragg, NC: JFK Special Warfare Center, August 1988), pp. 4-1 through 5-36.

[12] William Sweetman, *Stealth Aircraft* (Oscelo, WI: Motorbooks, 1985), pp. 78–82.

[13] *Ibid.,* pp. 46–52.

[14] Spokesman, Australian Defense Science and Technology Organization, as quoted in *Jane's Defence Weekly* (December 17, 1988): 1549.

[15] Ross S. Kelly, "N.A.T.O.'s Special Operations Forces," *Defense and Foreign Affairs* (February 1985): 32–33.

[16] Viktor Suvorov, *Inside Soviet Military Intelligence* (New York: Macmillan Publishing Co., 1984), pp. 140, 149. John Erickson et al., *Soviet Ground Forces: An Operational Assessment* (Boulder, CO: Westview Press, 1986), p. 197.

[17] Victoria Pope, "Soviet Bloc Troops Prowl in West Europe," *Wall Street Journal,* March 7, 1988, p. 14.

[18] Remarks on the AN-2 by Colonel William Lacey, Deputy Director for the U.S. Department of Defense Office of Operational Testing and Evaluation, in Debra Polsky, "A Red Army Invades New Mexico Desert for Pentagon Testing," *Defense News* (September 26, 1988): 3.

[19] Andrew W. Marshall and Charles Wolf, "Sources of Change in the Future Security Environment" (Washington, D.C.: Department of Defense, Commission on Integrated Long-Term Strategy, 1988), p. 19.

CHAPTER FIVE

Special Operations in Mid-Intensity Conflict

The vast majority of the world's military expenditures are directed at creating, maintaining, or improving conventionally organized armed forces. These military and naval forces are normally equipped with combat, transport, and reconnaissance aircraft; warships and naval support craft; artillery, trucks, tanks, and helicopters. Conventional formations are largely intended to achieve political goals by preparation for or the prosecution of mid-intensity conflict. Since the end of World War II, conventional forces have largely been unable to effect much political change. The Korean War, the Iran-Iraq War, and the Indo-Pakistani War resulted in stalemates. The various Arab-Israeli conflicts, as well as the Falklands War, closed with the original disputes firmly intact. The vast majority of the world's armed forces are organized, equipped, and trained to conduct mechanized warfare, and their ability to do so increases each year. The Soviet Union alone exported an annual average of almost 3,000 armored combat vehicles and 250 combat aircraft during the 1980s.[1] Mechanized warfare was conceived in order to break the tactical deadlock of World War I, disrupt linear combat, and swiftly bring about military conquest. Ironically, post–World War II mechanized forces have usually resulted in a condition of the political *status quo*.

Significantly, in one of the few late 20th Century conflicts that brought about political change, the victor avoided the use of a mechanized force structure for the great majority of the conflict. The Vietnamese communists weakened their adversaries by the employment of foot-mobile troops throughout the countryside, using depth to nullify their opponents' mobility, robbing them of the advantages of rapid concentration.

By introducing depth to the battlefield, they forced their mechanized opponents to attempt the impossible—defending everywhere and everything. Only in the very last stages of its long war did Hanoi use mechanized forces, and even then these elements were employed in close coordination with dismounted irregular forces operating behind the opponent's mechanized forces.[2]

The fact is that mechanized forces, originally created to overcome a linear battlefield, are themselves highly dependent on the existence of well-defended front lines. The need for lines is not surprising. Mechanized warfare depends on air power and, of course, airfields must be defended. Mechanized warfare depends on a constant supply of petroleum products, and the steady stream of fuel transporters must be protected. Mechanized warfare depends on functioning machines, so there must be secure areas, safe havens for mechanics, spare parts, and maintenance facilities. Mechanized warfare is dependent on the timely delivery of enormous tonnages, vast stocks of supplies that must be shipped, handled, stocked, moved, and protected. The conduct of mechanized warfare requires a protected rear, and the establishment of defensive lines is the most practical way to provide that. Mechanized warfare is inherently linear.

Mechanized warfare is slow. In part, this fact stems from the requirement for a defended line. Troops who have few other duties than to constitute that line must be deployed and supported. A Soviet study of the Red Army's World War II experience indicated that even during the height of Moscow's offensives, 84 to 93 percent of its front lines were on the defensive, protecting the shoulders and rear of attacking columns.[3] If a breakthrough is achieved, most commanders are reluctant to press on to the enemy rear if the cost is to expose their own rear areas to the opponent's mechanized forces. They are apt to rein in lead elements, consolidate gains, and move their all-important defensive lines forward in a ponderous advance before attempting another breakthrough.

Because of this, mechanized warfare has usually failed to match the rates of advance achieved by the Imperial German Army in its infantry attack during the Aisne Offensive of May 1918. Ludendorff's storm troopers, infiltrating French and British lines in small groups, averaged 18 kilometers per day. Erwin Rommel equaled that in his best offensive campaign in the Western Desert during World War II. Hitler's armies averaged only 10 kilometers per day against France in 1940, and 5 kilometers per day against the Soviets a year later. In their 900-day counteroffensive, the Soviets averaged 2.2 kilometers per day. Their British and American allies did a bit better in the west, 5 kilometers per day. The Israelis achieved 60 kilometers per day in the Six Day War but, when their enemies achieved a better state of organization, they managed only 5.6 kilometers per day in the Yom Kippur War, and just one kilometer per day in their 1982 incursion into Lebanon.[4]

The lack of speed on a mechanized battlefield is in large part due to weapons effectiveness. The World War I division could normally dominate a 10-square-kilometer area. Today's organization, with approximately the same manpower, can impede, delay, or perhaps defeat a similarly organized opponent's attack in a 40-square-kilometer area.[5] This area is likely to grow, since the one trend that is virtually certain is that weapons effectiveness will continue to increase.

The lack of battlefield speed is compounded by a lack of strategic speed, a particular problem for U.S. armed forces and N.A.T.O. The crux of this problem has to do with the growth in the sheer weight of America's ground forces, specifically the heavily mechanized U.S. Army, and the steadily diminishing availability of usable shipping. A U.S. reinforcement of Western Europe would have to be about 90 percent dependent on surface shipping. Current and planned airlift is simply inadequate. Despite an emphasis on light infantry units during the 1980s, the weight of an average U.S. Army division grew by about 19 percent from 1980 to 1989.[6] During the same period, U.S. commercial ocean-going shipbuilding facilities dropped from nineteen to only nine. In 1980, there were 142 such ships under construction in America. By 1988, none was being built. The U.S. flag merchant fleet is expected to drop from about 400 ships in the late 1980s to 200 by the year 2000.[7] An American commission, appointed by the president to study the problem, estimated the military sealift requirement at the turn of the century would be about 650 ships.[8] For all intents and purposes, the U.S. Army is becoming strategically immovable.

The air aspects of mechanized warfare have taken on the same characteristics as those on the ground. Although combat aircraft were capable of roaming over many of the World War II battlefields, that freedom was greatly reduced thirty years later. The United States and Israel, two nations enthusiastically devoted to air power, found extreme difficulty in penetrating air defense systems found in North Vietnam and those encountered during the Arab-Israeli War of 1973, respectively. The combination of mobile radar systems and accurate air defense missiles has made air attack about as costly and indecisive as the companion ground effort.

Left to their own devices, the advocates of mechanized forces plot a certain course to stalemate. This is perhaps a hopeful indication, at least for those nations that are dedicated to political *status quo*. However, the most consistent effort in military affairs is an unending search for advantage. The mechanized stalemate offers several opportunities for those who seek advantage, and at least two of these opportunities lie in fields that can be developed and exploited by special operations forces.

Given some initiative and imagination, special operations forces can have a considerable near-term impact on a mechanized battlefield, specifically in causing damage in an opponent's vulnerable and vital rear areas. In the more distant future, these forces may form the cadre to create a new type of defense system against mechanized forces, one that may avoid ever more costly investment in the arms race. Both of these innovations are, however, dependent on the skill and competence of the officers and soldiers who form the ranks of special operations forces.

A New Human Dimension

Western special operations forces are likely to be better led than their conventional counterparts in the late 1990s and early 21st Century. Few factors are more important to the success or failure of a military force than leadership. No matter how capable tomorrow's weapons systems may be, they will perform no better than the men who use them. With so much at stake, it would appear that the selection and development of leaders would receive as much investigation and attention as selection and development of weapons systems. That, however, is not the case for the vast majority of military organizations. Despite considerable societal change, the fundamental criterion for officer selection is much the same as it was in World War I. Leader selection is possibly the most tradition-bound of military activities. Fortunately, leader selection in some Western special operations forces has taken a basic departure from tradition.

Analyzing the Allied failure at Caporetto in 1917, most observers cited a systemic failure in Italian leadership. One perceptive analyst extended his investigation into the origins of Western practices for selecting military leaders. Gioacchino Volpe noted that the fundamental criterion for an officer's commission was the possession of a college degree. He found that this practice was based on several widely accepted notions. The university graduate, normally a member of the upper classes, could be assumed to have a higher stake in the survival of the state, the institution that provided or secured his position of privilege. Second, Volpe discovered that European upper classes were considered to be composed of families having a heritage of the warrior spirit. Additionally, it was widely assumed that the university prepared the graduate in technological fields, providing the state a class of citizen who would be capable of handling the complicated machines of modern warfare. Volpe's investigation revealed that the college graduate was believed to be accustomed to team sports, activities rarely practiced by early 20th Century lower classes, and that teamwork was known to be an important facet of modern armies. Finally, he discovered that upper-class university men were normally thought to be accustomed to handling servants, experience believed useful to the art of command.[9]

The university degree and officership are still linked in most military forces today, but there does not seem to be much basis for continuing this tradition. Even if all of the original assumptions for choosing college graduates as leaders were valid, those reasons have long since been overcome by social change. By and large, servants have disappeared. Team sports are no longer a requirement for most university students, and it is often the lower tiers of the Western socio-economic structure that excel in them. Increasingly, Western students avoid technological subjects in higher education, and it is again the lower social classes that are more apt to be familiar with machinery and electronics. The notion of an upper-class warrior spirit was probably spurious in the 19th Century, let alone the 20th. Finally, Western societies have long since grown accustomed to political dissatisfaction arising from their universities prior to similar sentiment being expressed by the public at large. Today, a university graduate may defend his society, but his political reliability can hardly be assumed to be superior to that of others. The idea of relating military leadership to the college degree is an idea whose time is past. It is one of the few Western practices that the Israeli Armed Forces, for example, have studiously avoided.

This is not to contend that military services are somehow inferior to civilian institutions in developing capable senior managers. There is evidence to the contrary. A ten-year study comparing test results of hundreds of American general officers and their counterparts in the U.S. business world showed the generals to be in the nation's top 95th IQ percentile, better educated, more self-assured, and having a higher degree of psychological health than their civilian equals.[10] But, the requirements of the commercial world are quite different from those of a military service, an institution whose ultimate worth must be demonstrated on a battlefield, not in the marketplace.

Leader selection and development in Western special operations forces began a departure from military norms after a perception of battlefield failure during the Malayan Emergency in the 1950s. The leadership of the SAS, dissatisfied with the unit's performance against communist terrorist bands, determined that a revision of the induction and initial training of SAS personnel was warranted. The program that was developed not only applied to the enlisted ranks; officers were also included in a demanding and wholly new selection process.

The SAS selection system eliminates candidates who are physically inferior, cannot exhibit sound independent judgment under stress, and lack determination. The system involves several weeks of arduous, individual land navigation treks. The candidates must carry heavy rucksacks. Each man plots his own lonely course day after day and cannot rely on others to make decisions. During the trial, candidates are not encouraged, but instead are given every opportunity to drop out of the course, an action that would eliminate their chances to join the unit. Normally, only about 15 to 25 percent of candidates are able to complete the course and be selected for membership in the regiment. The qualities of those who pass the trial include a high IQ, superb physical condition, and demonstrated ability to choose wisely despite conditions of great fatigue and mental stress. Only the determined, self-reliant, and quick-witted are selected to serve in the SAS.[11]

By creating a new selection process for all ranks, the SAS automatically revised its criteria for leader selection. It is perhaps not surprising that the failure rate of officers, leaders who have been successful in conventional British military units, is about the same as the failure rate for the enlisted ranks. In other words, qualities that satisfy most military leadership requirements in the United Kingdom do not suffice for leadership positions in this special operations unit. Since the end of the Malayan Emergency of the 1950s, the SAS has compiled an almost unbroken record of success, a record that has captured American interest and envy to such an extent that at least two U.S. special operations organizations have duplicated much of the SAS selection system.

During the late 1970s and early 1980s, the ground component of the U.S. special operations forces began a radical revision of its leadership selection and development procedures. The Delta Force, the American counterterrorist and commando unit, adopted the SAS selection process in 1978. U.S. Army Special Forces instituted a similar program ten years later. U.S. Army Ranger units were authorized to draw officer leaders from conventional units, selecting men who had already successfully demonstrated a talent for command at the same levels to which they were being assigned in the Ranger organization. Formalized personnel career fields were instituted for Special Forces, Civil Affairs, and Psychological Operations officers; these officers have thus far largely outpaced others in passing through career advancement hurdles such as promotion and school selection, a complete reversal of the American special operations officer experience in the 1960s and 1970s. There is every reason to expect that U.S. special operations ground elements, organizations that contain the great bulk of the American special operations capability, will in the future be far better led than they have ever been before.[12]

Both British and American special operations forces have now developed procedures to field intelligent, physically fit, determined, and self-reliant leaders on future battlefields. Because the selection process requires many to be tested for the few selected, it is expensive; however, few other actions could be so fundamental to success in combat. These units have departed from the leader selection procedures of most other military organizations, and it is quite likely that these special operations formations will be fully capable of outwitting and outlasting their adversaries. Not much is known about the leadership selection system in Spetsnaz, but it is believed to consist of only a few interviews.[13] The British and American systems have every promise of producing unprecedented levels of battlefield performance. And some of the future special operations tasks will undoubtedly demand previously unknown levels of achievement.

Reconnaissance and Target Designation

For a variety of reasons, mid-intensity conflict will probably continue to be characterized by mechanized warfare all through the 1990s, and special operations forces will in no way be decisive on these battlefields. The sheer magnitude of the inventories of mechanized equipment, combat aircraft, tanks, and artillery ensures that mechanized warfare will be the likely method of waging armed conflict between regularly organized military forces. Existing systems of training, force structure, and doctrine all point in this direction. Special operations forces cannot be expected to be determining factors in this type of conflict, simply because their numbers will continue to constitute only a small percentage of any armed force. Additionally, technological trends do not indicate that truly effective hand-held anti-tank weapons systems will be widely available until the turn of the century.

But, battlefield stalemates or even slow-moving fronts provide the opportunity to employ forces within an enemy's vulnerable rear areas, forces that can significantly contribute to a successful outcome by gaining information and inflicting damage on vital installations and the opponent's supply and maintenance lifelines. Small groups of eight to twelve men, capable of infiltrating into an enemy's rear, are difficult to identify. Once they gain those rear areas, they have the advantage of a low signature. Free of dependence on motors, propulsion systems, or any other emitting device except a radio, and free of the possession of any large object that can be visually or electronically identified at a distance, they can roam through the largely undefended rear areas with minimal risk. Their presence there adds depth to the battlefield. Survival for these groups will depend mostly on their training. If they attempt to attack targets themselves, they are likely to be hunted down, albeit at some cost to the defender. If they merely report enemy installations, units, and headquarters, their continued employment is more assured. They may also assume a target designation role and even provide terminal guidance to friendly weapons without substantial risk.

The growing need for special operations reconnaissance and target designation capabilities springs in part from widespread concern over the limited number and vulnerability of satellites.[14] Also there is a decreasing survivability of manned air reconnaissance vehicles. These concerns provided a stimulus for a massive $4 billion U.S. program to create the Joint Surveillance Target Attack Radar System (JSTARS), which is to consist of twenty-two E-8A aircraft equipped with powerful synthetic aperture and side-looking phased-array radar systems. Each aircraft will detect ground movement and even stationary targets over a 480-square-kilometer area in a region that is at a considerable distance from the plane. JSTARS is thus designed to fly over the assumed friendly skies, peering into enemy territory. The idea is immediately to target the opposition's forces with missiles or, in the event they are at shorter ranges, with the Multiple Launch Rocket System (MLRS).[15]

Unfortunately, JSTARS has limitations. The slant ranges of air defense systems like the Patriot and similar Soviet versions are growing so rapidly they already overlap enemy-held territory. That, of course, pushes JSTARS aircraft further back, for they must fly at high altitudes to achieve depth to the area that they are expected to scan, and these aircraft will be highly identifiable. Additionally, with so few aircraft for the system, it is certain they will be high on the list for early destruction by any opponent. JSTARS aircraft will probably be subjected to carefully planned raids by interceptors armed with long-range air-to-air missiles. JSTARS is an important addition to the Western arsenal, but it is not a system to be relied on exclusively.

With satellite surveillance and air reconnaissance in jeopardy, it is important to provide a supplement for (or even an alternative to) JSTARS. Otherwise, it is quite possible that future commanders will be ignorant of the enemy's rear area dispositions. Additionally, the same reasons that make for a diminished air reconnaissance capability also produce a lessened use of air attack in an opponent's rear. Against a substantial, well-deployed air defense, an air attack cannot be conducted if pilots have to search out their targets. They must know in advance where targets are so they can reduce their exposure to enemy air defenses. The clear need is for a capability to place the enemy rear under surveillance and conduct strikes as much as possible with accurate missiles.

Special operations surveillance and selective interdiction capabilities may be gained through careful planning, infiltration, and placement of well-trained and well-led teams. The teams must use clandestine communications systems, systems with a minimal chance of being detected. Periodically, the teams must be resupplied, possibly via cruise missiles. The surveillance portion of these missions has a primary goal of providing the friendly commander a steady stream of accurate ground-observed information on enemy dispositions and movement deep in an enemy's rear area. The selective interdiction portion of such operations has the goal of systematic destruction of some segment of the opponent's force structure or some category of his material needs. For example, special operations teams may be tasked to concentrate their efforts on the location of enemy air defense units and to provide terminal guidance for friendly missile strikes against air defense facilities. Or, they may be directed to focus on an opponent's ammunition supply and storage chain. The concept of selective interdiction is to use scarce resources wisely, concentrating on a well-chosen, critical element that would cripple an opponent rather than attempting a broad, unfocused attack on the rear areas.

Planning for surveillance and selective interdiction missions initially centers on establishing and maintaining multiple areas of operation along vital enemy supply and movement corridors. There may be as many as four or five teams placed along a corridor, each team being assigned an exclusive area that contains a source of fresh water and sufficient space and foliage or cover to ensure wide choices in hiding places and room for the teams to escape detection. It is important during the planning process that the teams be isolated from one another so that, if one team is eventually captured, its members cannot compromise the others.[16]

Team infiltration to the enemy rear will probably be accomplished by air transport, but the use of parachutes should be avoided. Against a well-equipped enemy, the final leg of infiltration may require the team to make extended ground movement on foot. Vital supply corridors will most probably be under the protective umbrella of an air defense system, so the surveillance and interdiction team may have to be initially landed at some distance from its designated area of operations and only arrive there after a series of night marches. Despite careful avoidance of major areas of enemy air defense strength, successful air infiltration will probably call for the use of deception and supporting air operations designed to decoy the opponent's radar and interception capabilities away from the actual delivery aircraft. The parachute drop is, of course, a standard delivery method, but it should be rejected if the use of short takeoff and landing (STOL) aircraft, tilt-rotor aircraft, or helicopters is feasible.

There are two good reasons to reduce reliance on parachute infiltration in the 1990s and beyond. First, the probable requirement for extensive ground movement after the initial drop-off places a premium on avoiding parachute-related injuries. Second, a parachute jump requires aircraft to achieve altitudes of 400 feet or so at the moment parachutists exit the plane so that parachute canopies can be fully deployed and stabilized prior to the parachutists' reaching the ground. Penetration aircraft can and often do fly at much lower altitudes. For example, the difference between 100 feet and 300 feet in altitude amounts to about nine nautical miles difference in ground radar intercept ranges. But, at the critical moment when the aircraft is over the drop zone, it may well have to rise to a radar identification zone, pinpointing the team's landing area.

On the other hand, a STOL, tilt-rotor aircraft, or helicopter is able to make its entire penetration run at lower altitudes than current or planned special operations fixed-wing aircraft. When on-board sensors give warning that the aircraft is being "painted" by either airborne or ground-based hostile radar, the STOL, helicopter, or tilt-rotor pilot has the option of immediately landing. The team can then be discharged, or the air crew and team can decide to simply wait for an hour or so before continuing their penetration mission. Until stealth air transports are available, parachute infiltration should be the exception and not the rule.

After arrival in the area of operations, the team's first order of business is to take in supplies and stock its assigned region. The purpose is to establish several hidden caches of food, ammunition, medical supplies, additional radios, and batteries. These caches will be used to sustain the team for as much as sixty days between supply deliveries. They also serve as a rallying point should enemy security forces cause the team to split up to evade pursuers. In a heavy air defense environment, these supplies must be brought to the team by cruise missile, and it may be necessary to launch several decoy missiles along with each missile supply mission in order to confuse and deceive hostile air defense units.

The team then begins its productive work. Moving with stealth and primarily at night, its members must avoid contact with enemy forces, but locate and identify enemy units and installations. Communicating once or twice a day, the team must use low-power meteor-burst transmissions or any other type of communications system that will provide the best chances of avoiding detection by hostile communications intercept units. However, the team must assume its transmissions are being intercepted and never transmit from the same spot twice, never establish a geographic or time-related pattern of communications transmissions, and never remain long in an area from which it transmits.

Target destruction should be accomplished by cruise missiles, either by simply conducting a strike on a team-reported location or by a coordinated effort by the team and the missile launch unit. The latter type of mission provides superior accuracy and is best achieved by the team's use of a laser designator. The designator is used by the team to "paint" the target shortly before a missile's scheduled arrival. Greater security for the team during strikes that require a high degree of accuracy can be gained by using a remotely controled laser designator. The team simply stakes out the designator, aims it at the intended target, and turns on the designator's receiver. The incoming missile must be transmitting on the desired frequency so that it will activate the designator's laser beam. The team can then be miles away, perhaps staking out another designator.

While cruise missiles are particularly useful for deep targets, there are other, shorter-range weapons systems that will be able to supplement these long-range missiles. The U.S. Army's Tactical Missile system may replace the older, 130-kilometer-range Lance missile in the mid-1990s. The new missile, already successfully flight tested, will carry anti-personnel, anti-tank, and "smart" warheads, the last capable of tracking and destroying moving ground targets. The thirty-kilometer-plus range U.S. Multiple Launch Rocket System should be completely fielded by 2010 and will deliver thermal guidance warheads, German-made scatterable anti-tank mines, and self-forging munitions. For this system, the self-forging anti-material munitions will be packaged six per rocket warhead.[17]

Surveillance and selective interdiction tasks accomplished by special operations forces can seriously damage a mechanized force. And, this function is particularly vital in an age when overhead reconnaissance and overflight of an enemy's rear are difficult, if not impossible. Realization of a potent special operations surveillance and selective interdiction capability calls for modification of some existing cruise missiles to handle the resupply and terminal guidance requirements. It also calls for additional peacetime radar mapping of potential conflict areas, work that probably should proceed in any event. It is quite possible that the costs would be far less than the $4 billion JSTARS program.

People's Armies

Almost every military institution represents itself as the protector of a nation's people and, in fact, many armies use the term *people's army* in their official titles, particularly armies of Marxist states. That term evokes the vision of a nation at arms and citizens defending home and hearth. The facts often tell a different story. Many of the poorest of nations have now acquired sophisticated war machinery—tanks, planes, and warships—all in the midst of squalor and poverty. The armies of these Third World states do not represent the technological levels of their populations, but actually reflect a level of material modernity far above the status of the peasantry they supposedly serve. Armies are overwhelmingly composed of the young men of a nation, and thus are not demographic cross-sections of a particular country. Additionally, the long training periods required to learn to operate the complicated machines and weapons of war have now pressed military institutions into a growing reliance on professionals. The great majority of all armies are male, mechanized, and professional and, as such, they are often societies apart, elite subcultures within a general population. They are not citizen armies.

However, citizen armies, or people's armies, do exist. For example, the Swiss have only about 1,500 professional soldiers, but they can field a 1.1-million-man, well-armed militia within twenty-four hours.[18] The Yugoslav Territorial Defense Force (TDF) provides a modern example of this type of organization. From the end of World War II until the late 1960s, Yugoslavia maintained a typical European military structure, a mechanized armed force. But the 1968 Soviet invasion of Czechoslovakia provoked a radical change in Belgrade's thinking. Realizing their country could never afford a mechanized force structure adequate to repel a Soviet invasion, Yugoslav leaders decided to augment their defense establishment with a massive paramilitary organization composed of citizen soldiers. Retaining its regular army, navy, and air force, but keeping them small, Belgrade's decision for a citizen army represented no offensive threat to its three Warsaw Pact neighbors, since the new program was purely defensive in character. The Territorial Defense Force does not have the transportation, communications system, logistical organization, battlefield mobility, or weaponry for cross-border offensive operations. It is an organization designed to head for the local hills at the outset of war, and one that will conduct operations in coordination with the country's regular forces. Or, if Soviet mechanized forces were to overrun portions of Yugoslavia, TDF units would remain near their villages and urban areas, within reach of hidden weapons caches and operating at night as guerrillas.

The Yugoslav Territorial Defense Force strength is nearing three million, about 15 percent of the population, but as training for this citizen army is rudimentary and its equipment needs minimal, the drain on the national treasury is quite small. In effect, TDF members contribute far more to Yugoslavia's GNP than they take from it. However, the country is somewhat unsuited for this type of force, since ethnic schisms limit the degree to which citizens' access to arms can be safely granted by the central government. For example, TDF members do not keep their weapons at home in the Swiss militia tradition. But, strategic placement of armories in villages for the rural peasantry and in over 2,000 factories for workers ensures that the TDF can field half its strength within twenty-four hours and the rest in another day.

On first glance, the Yugoslav border may seem inviting to a potential invader, but it is the next step, control of the nation, that is sobering to any would-be conqueror. The 1,700-kilometer-long border with Warsaw Pact nations is far more extensive than the inter-German border, and much of Yugoslavia's eastern frontier offers excellent trafficability for mechanized forces. However, once in, the attacker must contend with the TDF in his rear. Yugoslav analysts claim that a hostile occupying force would have to maintain roughly nine soldiers per square kilometer to keep the nation under nominal control and, even then, the TDF would probably be winning in some regions. Although an invasion of Yugoslavia would offer Moscow quick access to the Adriatic and seriously jeopardize N.A.T.O.'s southern flank, the required two-million man Soviet counterinsurgency force and the probability of a long, indeterminate guerrilla war must give considerable pause in the Kremlin.[19] In fact, the Soviets may prefer a push through Germany against a mechanized force rather than face the prospect of war against this nation, a country that has developed an on-call "instant insurgency." Yugoslavia has found a potent deterrent without resorting to nuclear arms, large and costly mechanized forces, or measures that provoke its neighbors into an arms race. It may have serious internal problems, but it appears rather secure from external aggression.

A modified Yugoslav concept of territorial defense has been proposed for Western Europe. The germ of the proposal emerged in 1981 when Major General Franz Uhle-Wetter of the Federal Republic of Germany advocated a radical revision of N.A.T.O.'s ground force structure. Uhle-Wetter envisioned a half-and-half mix of mechanized forces and lightly armed heliborne infantry. The latter forces were to be initially targeted on Soviet Spetsnaz and subversion threats, but were subsequently to work in close harmony with friendly mechanized elements. In 1982 and again in 1985, Brigadier Richard E. Simpkin began taking the concept farther, the light heliborne elements being rejected in favor of special forces. The British armor expert also proposed a large militia force, similar to the Yugoslav TDF, and projected special forces to be the focal point of future land force structures, forces he claimed were destined to become decisive instruments of action at the operational and strategic levels of war.[20]

The reason special operations forces and territorial defense or citizen armies are conceptually melded together is that each can complement the other, achieving a combined strength greater than their sum. Locally assigned special operations units must act as a small, full-time cadre for widely dispersed militia organizations. The citizen soldier goes about his civil pursuits on a day-to-day basis and is only occasionally involved in training, perhaps for a week or so each year. Planning, militia training, and daily administration for the unit are handled by professionals, the special operations detachments. The latter should manage and constantly monitor a complex of hidden arms and supply caches. These detachments also establish and operate clandestine communications nets and advise and assist their militia charges during war.

The idea of a territorial defense manned by militia forces is an old one, a concept largely discarded with the advent of mechanized warfare. But technology, arms race competition, and the increasing cost burdens of mechanized forces are bringing about new conditions. Militia forces have been no answer to tanks and airplanes through most of the 20th Century, but that is changing. U.S. Department of Defense studies indicate that 90 percent of all tactical aircraft downed in combat from 1975 to 1985 were the victims of heat-seeking missiles, an increasing percentage being of the shoulder-fired type.[21] These man-portable anti-aircraft missiles now work and work well. There is a good chance that a reliable and effective light anti-tank weapon will be available at the turn of the century, if not sooner. When that occurs, mechanized forces can be destroyed bit by bit if they intrude into a nation that has organized and armed itself for guerrilla warfare. This type of defense poses no threat to a peaceful neighbor. This type of defense cannot provoke an arms race. And, this type of defense is inexpensive. It will be a logical choice for some nations in the early years of the 21st Century.

New Directions for Mid-Intensity Conflict

Mid-intensity conflict has been characterized by mechanized warfare for most of the 20th Century, but that era is ending. Even if the end is not brought about by the factors mentioned in these pages, the era of mechanized warfare cannot continue indefinitely. All eras eventually come to a close, the only question being when the end comes about. The close of one era ushers in another and, in the realm of warfare, the characteristics of the new era provide causes for the demise of the old one. Forecasting the new era is at best a hazardous enterprise. One useful approach to forecasting is trend analysis. So in determining the likely character of future mid-intensity conflict and the probable crossover time for the new era, it is important to study trends.

The most prominent trends indicating the end of mechanized warfare include the factors of cost and effectiveness. The average yearly rise in the cost of tanks, combat aircraft, helicopters, artillery, and warships outpaces the annual GNP advance of the great majority of the world's nations. Decision makers have therefore been reduced to spending the same amount for a lesser number of weapons, choosing to devote a greater annual amount of their nation's wealth to military hardware, or spending less and getting far less. Added to these rising costs of newer weapons is the growth in operations and support expenses. In the U.S. Army, operations and support costs for tanks and helicopters rose between 50 and 100 percent in changing from older systems to newer ones during the 1980s.[22] These rising costs are even more dramatically seen in the U.S. Air Force. In the mid-1950s, the air force inventory included 26,000 planes. By 1990, that number had fallen to 9,100. But, the inflation-adjusted operations and maintenance expense for the smaller number of aircraft was 6 percent higher than the expense of the 1950s.[23] Mechanized warfare is pricing itself out of the market.

Mechanized warfare is also becoming less effective in securing political goals. Created in part to break lines, it is heavily dependent on lines. Conceived to restore mobility to the battlefield, it is now consistently producing stalemates. Mechanized warfare is by now rarely a tool to effect political change. It is usually only useful to preserve the *status quo*. It is therefore likely that the new era of combat between regularly organized forces, mid-intensity conflict, will be one characterized by a system of fighting that at least promises an ability to bring about political change. The new era must have battlefield supremacy over the old. So it is also likely that the new era will feature a system of combat that is not linear, a system that will defeat mechanized forces by placing widely dispersed military elements in the rear of a mechanized opponent. In short, the new era will feature an ability to conquer an opponent and bring with it battlefields of great depth.

The world's special operations forces will be prominent in the new era of mid-intensity conflict, but possibly not to the degree predicted by Brigadier Simpkin. Special operations forces have been created and developed to add depth to the battlefield. They cannot long survive linear combat against regular forces and are not organized or equipped to be employed in such a fashion. By themselves, they cannot defeat mechanized forces, but they have the ability to assist those who can. They are not the wave of the future in mid-intensity conflict.

In the offense, special operations forces are most useful in the role of surveillance and selective interdiction. They may provide future commanders the only reliable means of information about the opponent's rear areas and the best capability to strike vital installations and enemy units in that region. However, in that role, they are merely exploiting a vulnerability of linear combat. In essence, they derive their offensive utility from a characteristic of mechanized warfare. Special operations forces hold little promise for a commander who seeks military conquest through their use alone.

In the defense, special operations forces are likely to achieve their greatest utility by acting as the cadre for a territorial defense force. They are ideally suited to be the professional core of a citizen army or militia. In this role, their future utility is more apparent. With the advent of a reliable, effective, shoulder-fired anti-tank weapon, the power of offensive mechanized armies will begin a rapid fall. Here again, it is not the special operations force that is the main feature. It is the promise of a widely scattered militia force defeating a mechanized army by the use of hand-held anti-aircraft and anti-tank weapons that makes the difference.

The evolution of mid-intensity conflict is reaching a point of wholly new directions. Technological trends indicate the availability of an effective, man-portable anti-tank weapon by the year 2000. But technology is not the only determinant of war. Perhaps it is the new system of leader selection, one now being used by special operations forces, that will be the most important factor in bringing about change. After all, it is man that produces, shapes, and uses technology.

Notes

[1] Frank C. Carlucci, *Soviet Military Power: An Assessment of the Threat* (Washington, D.C.: USGPO, 1988), p. 22.

[2] General Van Tien Dung, *Our Great Spring Victory: An Account of the Liberation of South Vietnam,* translated by John Sprangens (London: Review Press, 1977), pp. 103–105, 113–115, 122, 138.

[3] M.M. Kirgan, *The Fronts Attacked* (Moscow: Military Publishing House, 1987), p. 52.

[4] Chris Bellamy, *The Future of Land Warfare* (New York: St. Martin's Press, 1987), pp. 277–279.

[5] Ibid., pp. 274–277.

[6] Benjamin F. Schemmer, "Airlift, Sealift in Short Supply," *Armed Forces Journal International* (May 1986): 66–68.

[7] Admiral Carlisle A.H. Trost, Chief of Naval Operations, "SOS for the Merchant Marine," *Defense 89* (March 1989): 14–16.

[8] Andrew Rosenthal, "U.S. Panel Warns of Transport Ship Shortage," *New York Times,* February 19, 1989, p. 33.

[9] Gioacchino Volpe, *October 1917: Dall' Isonzo Al Piave* (Rome: Liberia D'Italia, 1928), pp. 68–69.

[10] Michael Satchell, "The Military's New Stars," *U.S. News and World Report* (April 18, 1988): 33–42.

[11] Tony Geraghty, *Inside the SAS* (New York: Ballantine Books, 1980), pp. 224–244.

[12] Colonel Charlie A. Beckwith and Donald Knox, *Delta Force* (New York: Harcourt Brace Jovanovich, 1983), pp. 130–132. Brigadier General James A. Guest, *Army Times* (August 29, 1989): 10. Also, see General Guest's article, "Special Forces Training: New Initiatives to Enhance the Force," *Special Warfare* (April 1988): 5–11.

[13] Viktor Suvorov, *Spetsnaz: The Inside Story of the Soviet Special Forces* (New York: W.W. Norton Co., 1987), pp. 76–79.

[14] Raymond Colladay, Director, Defense Advanced Projects Agency, as quoted by Bob Davis, "Fleet of Defense Satellites May Be Future Pearl Harbor," *Wall Street Journal,* February 9, 1989, p. 1.

[15] Kent S. Sanderson, "Joint Stars Looks Deep to Win," *Field Artillery* (February 1988): 25–27.

[16] Author's concept presented to the commanding general of the U.S. XVIII Airborne Corps, April 1976, at Ft. Bragg, North Carolina.

[17] " 'SADARM' for MLRS," *Armed Forces Journal International* (December 1988): 34. Also see *Field Artillery Journal* (December 1988): 56–57.

[18] Bellamy, *Future of Land Warfare,* p. 43.

[19] Author's interview with Colonel Vidosav Novovic, Yugoslav Army, Carlisle Barracks, Pennsylvania, March 3, 1988.

[20] Richard E. Simpkin, *Antitank: An Airmechanized Response to Armored Threats in the 90s* (London: Brassey's Defence Publishers Ltd., 1982). Also see the same author's *Race to the Swift: Thoughts on Twenty-first Century Warfare* (London: Brassey's Defence Publishers, 1985), pp. 289–303.

[21] Glenn W. Goodman, "Services Adapt Airborne EW to Cope with Missile Threats," *Armed Forces Journal International* (February 1989): 81.

[22] General Maxwell Thurman, Commanding General, U.S. Army Training and Doctrine Command, as quoted in the *Washington Post,* November 13, 1988, p. A1.

[23] Franklin C. Spinney, Office of the Secretary of Defense, "Shape Up and Fly Right," *Washington Post,* April 1, 1989, p. B1.

CHAPTER SIX

Low-Intensity Conflict: Terrorism and Counterterrorism

As the Marxist world of the 1960s and early 1970s was absorbed in supporting insurgencies, the so-called wars of national liberation, the Western world and its Third World allies were deeply involved in what could be labeled the *counterinsurgency era.* With all of the attention in this realm of conflict focused on either insurgency or counterinsurgency, the terminology of the era was sufficient. There was no need for a omnibus label such as *low-intensity conflict,* a term that would capture and categorize politically inspired violence short of armed conflict between regularly organized military forces. However, beginning in the late 1960s, a rapid rise in the incidence of political terrorism made the lower end of the overall conflict spectrum more complex. By 1975, the Western focus on terrorism had already equaled or exceeded the attention and concern devoted to insurgency; new terminology and a new intellectual framework seemed appropriate. Thus terrorism was the base cause in the categorization of warfare into three levels of conflict, the struggle to define each of these levels, and the resulting cottage industry in writing about all of this.

But, a hard-headed analysis might logically conclude that terrorism did not merit the juggling of terminology and that it was not worthy of the attention it received. There is nothing particularly new about terrorism. Modern terrorism has not been the cause of any state's downfall. And, at least in the United States, yearly mortality figures from freezing, flash floods, and lightning strikes each normally exceed annual American deaths resulting from terrorist incidents. One can easily reach the conclusion that terrorism is an old story, an ineffective means of bringing about political change and, in relative terms, a rather insignificant activity.

However, along with its overly proclaimed past, terrorism probably has a sensational future, because terrorists persist in being adaptable to changing conditions. In an irrational, largely uncoordinated endeavor, there has been a rational, progressive evolution. Without any apparent central orchestration of their efforts, terrorists have been able to observe each other's failed attempts, learn from successes, and choose the next steps with a remarkable degree of foresight. To some extent, that can be seen in the logical evolution of terrorist strategies.

In the 1950s and early 1960s, a number of terrorist groups unsuccessfully pursued what could be called a strategy of exhaustion. The idea was to overthrow a government by a protracted campaign of terrorism, a campaign designed to wear down the will and steadily deplete the resources of security forces. For example, Colonel George Grivas-Dighenis directed an effort that included the use of over 4,800 bombs in a long-term effort aimed at taking Cyprus from British control and delivering the island into the hands of Greece. He managed to cause about 600 deaths, but the stubborn British outlasted the energetic colonel. At about the same time, but on the other side of the globe, a small number of Chinese waged a similar campaign. Unable to attract enough support to conduct a broadly based rural insurgency, the communist terrorists of Malaya tried and failed to exhaust the British in a campaign of terror. The Tupamaros reached the same fate during the 1960s in Uruguay. With the exception of the age-old and bloody religious struggle in Northern Ireland, this first phase of modern terrorism came to a close in the late 1960s. Terrorists had learned that governments were not likely to fall because of persistent, protracted terrorist attacks.

As the first phase was ending, the next was being born. This time, terrorists lowered their sights. The goal became one of gaining international recognition of a cause. The strategy of recognition can be described as *coercive propaganda,* a systematic program of violence to focus public attention on a specific issue. This strategy ensures continuing publicity for a cause that would otherwise be submerged in the day-to-day workings of the international news media. The blossoming terrorist strategy of recognition coincided with the initial worldwide flood of television receivers. The coincidence was not accidental.

The Palestine Liberation Organization and its many splinter groups used the strategy of recognition for over two decades. Italian, French, and German terrorist organizations have also pursued this strategy with varying degrees of success, and the strategy has also been employed by government-sponsored terrorists. For example, in 1980, the hand of the government of Iraq was revealed in London's Princes Gate episode. The incident, a hostage barricade situation at the Iranian Embassy, was planned and executed to draw attention to the supposed plight of the

Arab peoples of Kuzestan, a province of Iran. The attack was timed to precede the invasion of Iran by Iraqi armed forces.

In a limited way, terrorism has also been adapted to other forms of low-intensity conflict. After the failure of five major Latin American rural guerrilla campaigns in the 1960s (Guatemala, Venezuela, Columbia, Peru, and Bolivia) and the collapse of a number of major urban terrorist efforts in the mid-1970s, a merger of the two methods was accomplished under the Sandinistas in the late 1970s. The idea here was to so stretch the security forces of a beleaguered government that it would finally fall. The Somoza regime in Nicaragua was unable to muster the manpower for a protracted defense of both the cities and the countryside. In 1979, after twenty-five years of failed attempts, a Marxist regime was forcibly installed on the mainland of the Western Hemisphere.

Innovation by Latin American terrorists has provided some groups an independent source of funding that makes them less likely to be dependent on state support and more capable of moving across international borders. In the mid-1980s, the U.S. Central Intelligence Agency discovered evidence linking several terrorist organizations with international drug dealers.[1] The high profits from drug traffic and the use of smuggling routes add a measure of flexibility to the operations of "narco terrorists," placing them in a unique and powerful position.

Terrorists have also adapted their tactics to the demands of the time. Their operational techniques include assassination, bombings, kidnapping, and hostage barricade forays. The latter tactic has many forms: skyjacking, building barricades, and the temporary seizure of passenger ships and trains. In the face of a growing number of hostage rescue teams and a steady movement of governments toward flat "no negotiations" policies, terrorists have gradually moved away from sensational hostage barricade tactics. In a world increasingly dependent on centralization, including nuclear power plants, giant air transports, huge manufacturing plants, and high-rise apartment buildings, spectacular events become more possible. For example, the once diverse electrical power grid in the United States has gradually become more centralized so that, by the 1980s, it was dependent on only 4,000 large capacity transformers. A coordinated attack on some of these transformers could black out one-third of the nation and produce an extended period of national electricity shortage. Statistically, the long-term tactical trend is clearly toward a greater reliance on the well-placed bomb, and the targets are becoming more rewarding for the terrorist.[2]

Despite the efforts of many governments over the past two decades—efforts aimed at identifying and destroying terrorist organizations—the terrorist has proven to be amazingly resilient. When he is threatened or countered, he adapts and changes course. Failed strategies are discarded and new ones adopted. Risky tactics are shunned in favor of more promising ones. There appears to be little doubt concerning the future of terrorism. It should easily find its way into the 21st Century.

Mass Casualty Terrorism

Several Western authorities and government agencies predict future terrorist use of nuclear, chemical, and possibly biological weapons. For example, production techniques for a nerve gas, VX, have been published, and this chemical agent is known to be in the military inventories of Iran and Iraq.[3] A common 55-gallon drum of this agent fitted with a dispersal charge could constitute a serious threat to a major city's most densely populated area. The container itself would not have to be attended. It could be fitted with a simple timer and an anti-tamper ignition system.

Perhaps the most serious future terrorist threat is the possible use of a nuclear weapon. Some observers claim that ten kilograms of plutonium oxide and a substantial amount of high explosive material can be turned into a twenty-kiloton nuclear weapon. A blast of that magnitude would equal the devastation experienced by the Japanese city of Nagasaki during World War II. Thus far, concern has been tempered with knowledge of plutonium's limited availability. But, that happy condition is disappearing every day as more and more nuclear power plants accumulate the material. By the year 2000, annual worldwide production will reach 150,000 kilograms. Third World plutonium production at the turn of the century is estimated at about 12,000 kilograms yearly.[4]

However, a nuclear weapon is not easy to produce, and it would probably take state support to arm a terrorist organization with one. The requirements for production of such a weapon include a large, clandestine facility, several million dollars, great skill in the use of explosives, high-quality machine tools, and a considerable amount of time. Without the support of some nation, a terrorist would likely turn to theft. But a more likely option is to choose a chemical or biological weapon, a weapon that could be used to threaten a city.

There have already been incidents in which terrorists have been linked with biological agents, and chemical weapons are relatively easy to produce. A 1980 raid on a terrorist's apartment in Paris yielded a laboratory designed to produce biotulinal toxin, one of the most deadly substances known to man. Salmonella typhi, the bacteria that causes typhus, is another serious threat. It is estimated that a single pound of the bacteria, added to a million-plus-gallon water reservoir, could produce fifty million lethal doses.[5] And, there are growing numbers of facilities that cultivate and experiment with highly toxic microorganisms. By the late 1980s, there were already eighteen known culture collection centers in the world.[6] Almost any pesticide plant can be converted to produce nerve gas or any other type of chemical agent. The effects of an accident at a pesticide plant have been tragically demonstrated. The 1984 Bhopal, India, accident at the Union Carbide plant cost 2,500 lives. Any chemical factory producing pesticides and using the batch manufacturing process can easily switch from pesticides to nerve agents such as Sarin in about twelve hours.[7]

Counterterrorism

With trends of terrorism pointing toward increasing flexibility, strategies of recognition or insurgency support, a tactical use of bombs, and the growth of materials needed to manufacture nuclear weapons, toxic chemicals, and lethal microorganisms, the prospects for the counterterrorist may look rather bleak. But, in that arena there are reasons for optimism. The granting of terrorist demands has declined, there is growing, effective international counterterrorist cooperation, Western military retaliation against a nation that sponsored terrorism is now an established fact, and there are some promising technological aids that may come to fruition in the future.

The numbers of international terrorist incidents reached a zenith in 1985 and began a decided downward turn. Growing at a rate of between 20 and 30 percent in some years, incidents numbered 700 in 1985 as compared to about 200 per year in the late 1960s. Then, there was a 6 percent decline in 1986 and another 6 percent decline during the next year. In 1986, that first year of incident recession, skyjacking figures tumbled to the lowest point in twenty years. The declining rate was particularly noticeable in Europe. There, 1987's terrorist incident numbers were 137, down from 218 just two years earlier. There was a slight increase later, and there may be other irregularities in the future, but the "profitability" of terrorism has diminished. The probability of a terrorist's gaining his demands declined 50 percent during the 1970s and continued to fall during the 1980s.[8]

International counterterrorist cooperation, mostly coming in the mid-1980s, has contributed to the identification, detention, and, on occasion, the arrest of terrorists.[9] The 1986 U.S. retaliatory raid on Libya is often cited as being the stimulus for this cooperation. European governments finally realized the serious nature of American concerns and began acting with greater effectiveness in the hope of averting the unpredictable results of more U.S. military reprisals. A European example of this cooperation occurred in 1988, shortly after the "City of Poros" incident in which terrorists attacked a Greek excursion vessel, killing nine and wounding eighty. By quickly exchanging intelligence information on fingerprints and weapons sales, the governments of Italy and Greece identified the terrorist group, its drug trade connections, and the possible involvement of Libya. Even old-time adversaries have at last hinted at cooperation against terrorists. The U.S. Central Intelligence Agency has reported some intimations of Soviet assistance in the late 1980s, and the matter was evidently discussed between the U.S. secretary of state and the Soviet foreign minister in the spring of 1989.[10]

A more concrete indication of the worth of international counterterrorist cooperation can be seen in the growing numbers of terrorist-related detentions and arrests made by the U.S. Immigration and Naturalization Service (INS). Using both the European-based INPOL, a computerized intelligence service, and other information, the INS increased its detentions and arrests of known and suspected terrorists from three in 1982 to six in 1983. In 1984, INS officers arrested or detained nine suspects, and in 1985 the count was twenty-three. Terrorist incidents in the United States declined from 112 in 1978 to only seven in 1987. The U.S. Federal Bureau of Investigation estimated it was successful in preventing forty-five terrorist incidents during the period 1983 through 1987.[11]

Information exchange is only one form of international cooperation to assist counterterrorist officers. Suspected terrorists find it increasingly difficult to travel from one country to another due to coordinated visa denials. Quiet agreements conducted between various foreign ministries have resulted in a number of nations' refusing a visa if an allied country does so on the basis of suspicion alone. Additionally, there has been a significant amount of cooperation in the disclosure of trans-national bank accounts. Where terrorism is at issue, the inviolable Swiss bank account is becoming more of a myth than a reality.

The technology race between the terrorist and the counterterrorist is one where the terrorist appears to have the advantage, but the margin is not wide. The chief edge the terrorist is likely to have is not solely restricted to the area of nuclear, biological, and chemical weapons. The age-old art of assassination may reach a new plateau in the late 1990s and early 21st Century. The prime means of killing a chief of state or other important personality is likely to be a shoulder-fired anti-aircraft missile. Competition between those who design airborne missile deception and decoy systems and those who design anti-aircraft missiles is likely to be continuous and well funded. Another assassination device, the laser rifle, poses yet another serious threat to the lives of high officials. These threats and others, such as plastic handguns, will probably result in a steady decline in personal appearances by public figures throughout the 1990s and early 21st Century, a phenomenon that will in itself be an achievement for the terrorist.

The counterterrorist will likely enjoy technological support from a number of quarters. The first major hardware advance will possibly come about in the mid-1990s, if not before. The U.S. State Department is already field testing a flashlight-sized device that promises detection of any explosive, concealed or not. Should it prove reliable, searches of automobiles and visitors at embassy sites will become far more effective. Another device, one that will probably be of greater utility, will be employed for screening airline passengers and their baggage.

Individuals can be checked for the possession of explosives, or even for their recent handling of explosives, by a vapor ionization process. The subject enters a chamber and is exposed to a rush of air. The system analyzes the air wash, identifying any traces of explosive material. This personnel check system is expected to be fielded by the U.S. Federal Aviation Agency in the mid-1990s. A more important explosive detection system, however, may be employed at airports by the agency in the early 1990s. Passenger baggage, sometimes used to introduce terrorist bombs on airliners, could be subjected to a thermal neutron analysis screening, effective in identifying even hermetically sealed explosives.[12] Pressure chambers are already in limited use to initiate barometric fuses prior to loading baggage aboard passenger aircraft.

Concerns of the late 20th Century are making it imperative to deny movement to some international travelers. Terrorism represents only one of these concerns. The highly lethal Acquired Immune Deficiency Syndrome (AIDS) epidemic demands action, and a logical preventive measure is to identify carriers and prohibit them from entering regions where effective prophylactic measures are in effect. In 1989, the Soviet Union began instituting blood test screening for those travelers who intended to remain in the U.S.S.R. longer than ninety days. Persons known or suspected to be engaged in either illegal drug use or drug sales are likewise candidates for prohibited or strictly controlled travel. Together with obvious counterterrorist needs, these concerns make it possible that 21st Century international travel may differ considerably from the relatively simple jaunts of the 20th Century.

Gaining a visa to enter a foreign country, when one is even required, is now often only a matter of mailing a passport to a consulate or embassy and awaiting a perfunctory records check. On arrival at the destination, the traveler is often subjected to little more than a likeness comparison between his face and the passport photo. In this outdated, faulty system, all manner of things can and often do go wrong. Illegal passports of convincing quality can be bought for a nominal fee. Once in the foreign country, or even prior to travel there, the identification photograph can be switched. There are many other dodges. Worldwide, harried customs officials are only able to detain or arrest those terrorist suspects, fugitives, or criminals who are foolishly reckless, or those who have been specifically identified and tracked beforehand, usually by a friendly foreign security service. The current "system" is only of utility in catching the stupid, the unlucky, or the betrayed.

Technology exists that can be used to precisely identify the traveler in two ways. A laser handprint scan can mathematically record the tiny variations of ridges, curls, and fold lines of the human palm and fingers, providing a unique personal identification. A pilot project, the Automated Fingerprint Identification System, was comparing a set of prints with 9,000 others in less than one minute in the late 1980s.[13] Although a practical model of the entire system does not exist, it is possible that one will be available by the turn of the century.

There is, however, a precise identification system in existence. Each human has a unique protein coding in the DNA molecule. This unique, individual code can be isolated in samples of a person's saliva, urine, bone, semen, skin, or blood. Developed in the United Kingdom, the technology is being rapidly exploited within the United States. By 1987, two American commercial laboratories specializing in such work were operating at full capacity, and the U.S. Federal Bureau of Investigation opened a third in 1988. In the same year, the Bureau of Forensic Services of the state of California established a DNA computerized data bank containing protein coding for over 4,000 released prisoners.[14]

A "DNA visa" offers a better system to control international travel. The traveler would have to present himself at the foreign embassy or consulate, have a pinprick blood sample taken, and await the necessary verification procedure. The receiving government would then electronically compare the traveler's protein code against records of known criminals, fugitives, or terrorists. The individual's code can be reduced to numbers and transmitted in a message, so there is no reason for a paper passport or identity photograph other than to document the native country's permission for one of its citizens to travel abroad, a verification that could again be accomplished by a blood sample and electronic comparison at the point of departure. On arrival, the traveler may again be subjected to a DNA comparison to ensure he is in fact the person who applied for entry and not someone else.

A DNA visa would be no defense against a newly recruited or previously unidentified terrorist, but it would provide a powerful investigative tool. Electronic comparisons of travelers' records whose visits coincided with terrorist events could yield the field of suspects. The adoption of this technique, however, is more likely to come about because of public health concerns, not terrorism.

Counterterrorism may also receive a substantial assist from biotechnology, but indirectly. If the U.S. government's efforts to halt the illegal narcotics trade continues to fail, biological warfare techniques may be used. And, in the late 1980s, international production of both opium and coca persists.[15] It is feasible to create and dispense blights to attack the opium poppy, coca, and marijuana plants. While that effort would be directed at reducing the American domestic drug problem, it would seriously affect the fortunes of narco-terrorists.

All told, there has been considerable success against terrorists, yet that success carries a germ of failure. Implementing programs to defeat terrorism requires public consent and encouragement, responses that weaken as the incidence rate of terrorism drops. The occasional dramatic terrorist event heightens public interest temporarily, but media attention quickly wanes. It is likely that only a massive terrorist incident would carry the day toward implementing the costly, restrictive measures essential to fully deal with the problem.

Incidental Benefits of Counterterrorism

Along with the many debits that terrorism carries with it, there are a few unintended credits, particularly in the West. One positive aspect has been a general acknowledgment of the need for military forces. Historically, protracted periods of peace in the West have spawned pacifist sentiment. However, modern terrorism is often associated with some form of national insult that engenders a fundamental public understanding of the need for instruments of force.

Secondly, the West has gained a rather well-coordinated peacetime intelligence capability that far exceeds what would be expected in otherwise quiet periods. There are literally thousands of Western intelligence agents and analysts devoted to discovering and tracking terrorist organizations, and a growing corps of public servants who contribute to campaigns against drug dealers and hostile states that support terrorism. The enhanced Western intelligence capability owes much to modern terrorism and constitutes an existing and prepared wartime asset. Another unintended benefit also contributes to Western war readiness: physical security of sensitive installations. Construction of security barriers and surveillance provisions are among the initial time-consuming and costly measures undertaken at the outset of a conflict, periods when democratic states should be devoting their energies in more productive directions. The terrorist has provoked the West into peacetime security programs that place it on a near-war footing.

Finally, thanks to terrorism, the West has been able to test its decision process for the use of force on an almost regular basis. The Western systems of checks and balances for employing force are notoriously inefficient when compared to the systems of totalitarian states. But, Western decision cycles do exist and, although approval for the use of force by a counterterrorist unit is rarely granted, the planning process is often tested. Because of terrorism, the West is a little better informed, more secure, and a bit more ready for war.

A Clandestine Art

Terrorism persists in part because of its record of flexibility, but it also persists because of secrecy. If terrorism is state-sponsored, it is certainly supported in secret. While the prevalent strategy of the terrorist is the strategy of recognition, some terrorist events are contrived so as to be entirely hidden from public view. The clandestine nature of terrorism is the first and most significant barrier to counterterrorist efforts.

One of the most productive and effective techniques that law enforcement agencies use against criminal groups is the employment of penetrators, agents who join an illegal group in order to gain information. But that tool can be made useless by a determined terrorist organization. For example, one of the initiation events for an Italian Red Brigade recruit was to murder an Italian policeman, a policeman selected at random by the organization's leadership. That procedure was effective in deterring Italian security services from employing penetrators, and it is a technique probably used by the more established terrorist groups. The murder is, of course, not heralded by the terrorist organization. It is simply recorded as another unexplained murder of a law enforcement officer. Clearly, penetration of terrorist organizations can be extremely difficult.

Determining external support and assistance to terrorist organizations is equally difficult. In the early 1980s, the Central Intelligence Agency was unable to verify the training and support given to terrorist organizations for a considerable length of time.[16] Due to channeling support and assistance through espionage organs, the complicity of a sponsoring nation may go undiscovered and undocumented for years.

Perhaps the most sinister and powerful use of terrorism is concealed kidnapping. This technique can contribute to the overall strategic goal of the terrorist with little risk. An example of concealed kidnapping was accidentally discovered in a Third World state, a national allied with the West in the early 1980s. An American counterterrorist team, dispatched to assist this nation during one terrorist incident, inadvertently found the existence of another terrorist event. The granddaughter of a local newspaper publisher had been kidnapped by a terrorist organization and held for over a month. The crime had gone unreported because the publisher feared for his granddaughter's life. The kidnappers were eventually killed and the child freed, but the *quid pro quo* demanded by the terrorists was never discovered.[17] Undoubtedly, the terrorists were demanding a change in the publisher's editorial policy and news coverage. This type of terrorist technique is likely to go unnoticed and unreported for lengthy periods, involves minimum risk for the terrorist, and can be used to focus public attention on a cause via clandestine coercion. It may well be in widespread use.

Defeating the Strategy of Recognition

The terrorist holds the initiative and there is nothing on the horizon that promises to alter that. The Western response is defensive and reactive. It seems there is nothing to attack other than the occasional and briefly exposed radical and violent group. In some rare instances, a government is linked to a terrorist organization. But, when one of these bands is crippled, another invariably pops up in its place. All of this is particularly frustrating in the West, where counterterrorism is often portrayed as a just crusade, a righteous war. Western notions about war are heavily influenced by Clausewitz or, more accurately, by his countless interpreters. For almost two centuries, the basic Western premise of war has been that all energies must be devoted to the defeat of the enemy's forces. But, with the mercurial nature of terrorist groups, it is impossible to defeat all of the world's terrorists. A part of the problem has to do with the way the West thinks about war.

Eastern thinking may be more appropriate to the war against terrorism. In the 4th Century before Christ, Sun Tzu staked out an analysis of war differing markedly from those of later Western theorists such as Clausewitz. While Clausewitz concentrated on the defeat of enemy forces, the Chinese thinker saw that as a last resort. He advocated the destruction of an enemy's alliances as the preferred strategy. Failing that, Sun Tzu recommended defeat of the enemy's strategy. Finally, if this second approach were not possible, it was then and only then that he advocated destruction of the enemy army.

Applying Sun Tzu's thinking to the war against terrorism, his preferred strategy, the destruction of alliances, is an unlikely candidate. The relationships between various terrorist groups and, in turn, the relationships between some of these groups and supporting nations are too obscure to present reasonable targets. The least preferred strategy, destruction of all the terrorist groups, is also of little help, as noted above. That leaves Sun Tzu's second preferred strategy, destruction of the opponent's strategy. But can the strategy of terrorism be defined? And can it be defeated?

The strategy of most terrorist organizations can be defined. In the last decade of the 20th Century, the overwhelming majority of terrorist groups are using the strategy of recognition. The exceptions are the terrorists of Northern Ireland—who might be pursuing the strategy of exhaustion—some state-supported terrorist organizations, and the odd terrorist group involved in the support of a rural insurgency. In essence, the strategy of recognition involves little more than the creation of violent acts to direct public attention at a specific cause. The strategy requires a continuous cycle of three sequential events: an outrageous act staged in a way that carries a message, a subsequent international news story about the act that conveys the message, and consideration of the message by an international audience. A short definition of the *strategy of recognition* is the rational use of violence to convey a message. Another definition was offered by the vice president of the United States, George Bush, in 1988. The future president called it ". . . a cheap and effective way to . . . hijack the international media."[18]

The strategy of recognition can be defeated. It cannot be defeated by an attempt to eliminate violence, however, because as long as repression and persecution are used or perceived to be used, violence will exist. It cannot be defeated by somehow ending the reporting of violence, because violence always attracts man's interest. The strategy of recognition can only be defeated by stripping out the terrorist's intended message in the inevitable report of his staged event.

The strategy used by most terrorist organizations is wholly dependent on conveying a message. Whether journalists or broadcasters like it or not, the relationship between the majority of terrorist organizations and the mass media is symbiotic.[19] But, for the modern terrorist, it is not enough for the international news media to report "A Middle East terrorist group kidnapped two and killed one today." There are many "Middle East terrorist groups." The report must at least carry the name of the terrorist group, and many of these groups have specifically chosen names that identify their cause. In kidnappings and hostage barricade events, the terrorist's cause is often further elaborated in the list of his demands. In the case of bombings, a growing terrorist tactic, news organizations often fail to report the names of terrorist victims. But these same organizations usually go to great pains in attempting to identify the terrorists and their cause, often giving free publicity to several terrorist groups in the process. The international media make the strategy of recognition work. But that strategy would not work if the name of the terrorist organization and its cause or its message were not conveyed. There would be little point in the violence and no reward for the terrorist's risk if there were no message.

Increasingly, there is concern over terrorists' use of the international media to achieve their ends. In a 1986 Gallup poll sampling American public opinion, a 51 percent majority of respondents agreed with the statement that the press gave too much coverage to terrorists. A larger percentage, 60 percent, believed that press treatment of terrorism increased the chances of future terrorist acts.[20] In the United Kingdom, formal attempts to restrict media coverage of terrorism came in 1988. In banning direct radio and television interviews with the spokesmen of outlawed groups, Britain's home secretary stated terrorists were drawing support and sustenance from their access to television.[21] And in a decided departure from previous policies, America's most prestigious newspaper, the *New York Times,* in 1989 editorialized against publication of terrorist threats.[22] However, without a great and tragic event there is little hope of truly effective measures being taken to block the terrorist's message.

21st Century Terrorism and Counterterrorism

The West may have less to fear from the nuclear armed Marxist bloc than it has to fear from a varied group of political zealots representing obscure causes. However, a declining incidence rate of international terrorism could lull Western authorities and Western peoples into a false sense of security. There will probably be little political support for media controls, and thus a continuation of the strategy of recognition by terrorist groups. Unless the AIDS epidemic reaches gigantic proportions, there is not likely to be much support for a DNA visa. The terrorism of the 1980s was not enough of a threat to cause such a drastic program. Narco-terrorism has not been enough of a menace to provoke the use of biological warfare techniques on drug cultivation areas. Only massive, tragic incidents would provoke such departure from Western norms.

In some respects, the West is weakened by its successes. The steady, albeit uneven, decline of terrorism is largely due to sound police work and international cooperation by counterterrorist forces. But that success constitutes no bar to a determined, well-led terrorist organization. Western success is highly dependent on threat perception and, if the threat declines, so will the defense. It will take only one major incident of nuclear, biological, or chemical terrorism to revive a high interest in counterterrorism, but by that time it may be too late for some Western city.

The pity is that a method to severely damage the fortunes of international terrorism through attacking its most typical strategy is at hand. A refusal to repeat the name of the terrorist organization or publish or broadcast any terrorist demand would rob most of the world's terrorists of their messages, their purposes, and their rewards.

Notes

[1] Director of Central Intelligence William J. Casey, "The International Linkages: What Do We Know?" in Uri Ra'anan et al., ed., *Hydra of Carnage: International Linkages of Terrorism, The Witnesses Speak* (Boston: D.C. Heath, 1986), p. 10.

[2] Charles H. White, National Electrical Manufacturers Association, in testimony before the Senate Governmental Affairs Committee, February 7, 1989, as quoted in the *New York Times,* February 8, 1989, p. A20. Deputy Assistant Secretary of Defense Noel C. Koch, "Terrorism: The Undeclared War," in *Defense 85* (Washington, D.C.: Department of Defense, 1985), pp. 7–13.

[3] Major William H. Thornton, *Modern Terrorism: The Potential for Increased Lethality* (Langley AFB, VA: Army–Air Force Center for Low-Intensity Conflict, 1987), p. 7.

[4] Frank Barnaby, *The Automated Battlefield* (New York: Free Press, 1986), pp. 132–136. Louis Rene Beres, "Is Nuclear Terrorism Plausible?" in Paul Leventhal and Yonah Alexander, eds., *Nuclear Terrorism: Defining the Threat* (McLean, VA: Pergamon-Brassey's, 1986), pp. 45–62.

[5] Dr. Erhard Geissler, Central Institute of Molecular Biology, German Democratic Republic, as quoted in the *Washington Post,* January 24, 1989, p. 7.

[6] Robert Stevenson, Director, American Type Culture Collection, as quoted in the *Washington Post,* January 29, 1989, p. A34.

[7] Neil C. Livingstone, President, Institute on Terrorism and Subnational Conflict, "The Impact of Technological Innovation" in Ra'anan et al., *Hydra of Carnage,* pp. 137–153. Lynn Hansen, Assistant Director, Bureau of Multilateral Negotiations, U.S. Arms Control and Disarmament Agency, as quoted in *Defense News* (February 13, 1989): 28.

[8] Director of U.S. State Department Office of Counterterrorism, Ambassador L. Paul Bremmer, as quoted in *The Christian Science Monitor,* December 29, 1987, p. 3. Brian M. Jenkins, *International Terrorism: The Other World War* (Santa Monica, CA: The Rand Corporation, 1985), p. 13.

[9] Robert Oakley, Director, U.S. State Department Office for Counterterrorism,

"Terrorism: The Fight," *The Army Quarterly and Defence Journal* (April 1985): 167–180.

[10] William Webster, Director of Central Intelligence, as quoted in *USA Today,* January 12, 1989, p. 7A. Steven Engelberg, "Soviets Agree to Discuss Terrorism," *New York Times,* April 7, 1989, p. A6.

[11] Robert A. Ricks, Inspector, Federal Bureau of Investigation, as quoted in his November 22, 1988 presentation at George Washington University, Washington, D.C.

[12] Dr. Scott A. McLuckey, Oak Ridge National Laboratory, as quoted in the *New York Times,* December 28, 1988, p. A3.

[13] Donald R. Belsole, Criminal Justice Director, State of New Jersey, as quoted in the *New York Times,* December 8, 1988, p. 37.

[14] FBI Director William S. Sessions, as quoted in Stephen G. Michaud, "DNA Detectives," *New York Times Magazine* (November 6, 1988): 70–104.

[15] Michael Isikoff, "Opium, Cocaine Crops Rose Sharply in 1988," *Washington Post,* March 2, 1989, A3.

[16] Deputy Director, Central Intelligence Agency, Robert M. Gates, "The CIA and American Foreign Policy," *Foreign Affairs* (Winter 87/88): 221.

[17] Personal experience of the author, 1981.

[18] George Bush, Vice President of the United States. *Terrorist Group Profiles* (Washington, D.C.: USGPO, 1988), p. ii.

[19] Noemi Gar-or, *International Cooperation to Suppress Terrorism* (New York: St. Martin's Press, 1985), pp. 15–17.

[20] *New York Times,* January 26, 1987, p. A36.

[21] U.K. Home Secretary Douglas Hurd, as quoted in the *New York Times,* October 20, 1988, p. A11.

[22] Editorial, *New York Times,* April 10, 1989, p. A16.

CHAPTER SEVEN

Low-Intensity Conflict: Insurgency and Counterinsurgency

A sea change—a difference in wind, temperature, and current—requires the sailor to trim his craft, adjust, and perhaps even change course. It is the same for nations. There was a "sea change" in the 1970s. Western counterinsurgency campaigns that had featured British, French, American, and Portuguese soldiers pitted against guerrilla forces began to die out. The campaigns had been bitterly fought, unpopular with Western voters, and largely unsuccessful. Even though these conflicts ended with reduced Western influence in the southern hemisphere and lower tier of the northern hemisphere, there was a general sense of relief when they were over. The withdrawal of Western military forces and their return to the democratic heartland was welcomed. A more tranquil world, more placid "seas," were eagerly anticipated. But, peace among the poorer peoples of the globe did not come.

As soon as the colonial and post-colonial wars ended, another genus of conflict began, wars that were unexpected by Western leaders. Some of these wars were located in the same regions from which Western troops had only recently departed. The characteristics of the new conflicts were almost identical to the previous ones, but the participants had changed. One group of distinguished Americans, long familiar with international security affairs, considered the new conditions and concluded that the original Western perceptions had been flawed. They said that these types of wars had been incorrectly viewed as "a succession of transient and isolated crises." They now saw interrelationships.[1] At about the same time, another U.S. study group concluded that the United States simply did not understand the phenomenon of low-intensity conflict.[2]

To the West, there seemed to be two waves of conflict in the south. The first series of wars occurred during the 1950s and 1960s. Another set of conflicts began in the 1970s and appears to be continuing. Considering the results of the first period, it is easy to conclude that the West had little understanding of the phenomenon, because Western armies often placed a poor second in those conflicts. Perhaps the currents of the time were against the West. How the West will fare in the midst of the second wave is uncertain. There is, however, an indication that the current has changed. It may now favor the course of the West.

Revolutionary Warfare

Modern insurgencies and the measures to defeat them, counterinsurgency campaigns, are almost wholly found in the Third World. Industrialized countries are not likely to experience armed uprisings, since these nations are either democracies, where votes are preferred to bullets, or totalitarian regimes. In the latter case, the industrialized totalitarian state rests on stringent population control measures that make an insurgency extremely difficult to begin or maintain. Yet, there have been uprisings within Second World totalitarian countries.

The first modern totalitarian regime, the Soviet Union, experienced insurgency in the Ukraine during World War II, and the first Fascist nation, Mussolini's Italy, had to contend with an armed uprising during the same period. Both of these insurgencies were externally nurtured, the former by Nazi Germany and the latter by Britain and the United States. Both insurgencies were minor events in a greater drama and were entirely contingent on the conditions of a global conflict. An attempt to foster insurgency within a powerful totalitarian state risks bringing on a general war, and such schemes have wisely been avoided. Thus, the modern industrialized totalitarian state has been relatively immune to insurgency during periods of general peace.

However, Third World clients of both industrialized democracies and totalitarian regimes have experienced insurgencies, many of them sponsored or supported by First and Second World nations. During the last half of the 20th Century, a complex set of unwritten rules of international conduct has evolved, permitting armed competition between industrialized states within the borders of their Third World friends and allies. In some respects, these conflicts are surrogate wars, taking the place of violent confrontations between the great powers. These wars often appear to allow armed contests between First and Second World ideologies with minimal risk to the more affluent peoples of the northern hemisphere.

Yet, appearances may be deceiving. Many Third World insurgents probably adopt the ideology of a powerful nation in order to attract support of a rich patron. The insurgent's real goal may be simply to gain power in his native land. What may appear as a chapter in a global political contest may actually only represent an incident in an age-old story: the "outs" versus the "ins," another civil war. It is therefore doubtful that an end to ideological competition would bring about an end to Third World insurgencies. Both insurgency and counterinsurgency have an all too promising future in the 21st Century.

Insurgency is armed revolution. The fundamental concept of insurgency is the use of native resources to overthrow a government. There is often some form of outside assistance, but excessive foreign support can be counterproductive to the insurgent's cause. An insurgent cannot risk being branded as the tool of external interests if he expects and needs local enthusiasm for his cause. The insurgent's actions must therefore appear to be the natural and following reaction to local repression and injustice. All of this is essential because the success of an insurgency is highly dependent on the effective prosecution of revolutionary warfare.

The term *revolutionary warfare* has two meanings. Obviously, it incorporates the idea of revolution, the overthrow of a government. But, it has a second meaning that connotes the new, or novel. In this sense, revolutionary warfare conveys a European description of post–World War II anti-colonial wars in which Third World political movements successfully waged several campaigns to destroy colonial regimes. These insurgent movements were often divided into three interdependent components: guerrilla units, the auxiliary, and the underground.

The most visible component of a successful insurgent organization, the guerrilla force, often comprises only about 10 percent of the overall movement. In the initial stages of an insurgency, guerrillas may spend most of their time pursuing their civil occupations, fighting rarely and, even then, mostly at night. Organizationally based on the village or district, small units of ten to twenty guerrillas may begin consolidation in the latter phases of an insurgency so that company- and battalion-sized formations conduct stand-up battle with government counterinsurgency forces. At this stage, former guerrilla units may be fighting on a full-time basis, and the resulting war could be accurately characterized as mid-intensity conflict, combat between regularly organized military forces.

The largest and perhaps most important component of an insurgent organization is the auxiliary. Composed of unarmed supporters of the guerrillas, this politicized body of citizens may make up 80 to 85 percent of an insurgent movement. Like the guerrilla structure, the auxiliary is formed into locally based clandestine organizations, groups that supply recruits, clothing, shelter, food, and intelligence to the guerrillas. The auxiliaries also perform a vital counterintelligence function, preventing government penetration agents from joining guerrilla units.

The third clandestine component of an insurgent structure, the underground, often contains the "brain" of the movement, the political and military headquarters. Usually located within an urban area, the underground consists of intelligence and counterintelligence elements, perhaps a political council, couriers, and possibly terrorist cells. The underground not only sets the political agenda for the movement, it orchestrates the efforts of the guerrillas and auxiliary, designs campaigns, and directs the propaganda program.

This type of insurgent structure, familiar enough in the late 20th Century, was a novel creation in the Western world of the mid-part of the century and persists as something of a mystery in the West. While Western states waged guerrilla campaigns against Fascist regimes during World War II, they did so only as an adjunct to conventional war, using friendly guerrillas to harass and disrupt their enemies in order to make the tasks of regular forces a bit easier to accomplish. The surprise to many Western authorities following World War II was that "guerrilla warfare" would not only be used against the West in colonial lands, but used as a comprehensive form of warfare, independent of external regular forces. Therefore, the type of conflict waged by these insurgent organizations earned a special Western label, "revolutionary warfare."

During the third quarter of the 20th Century, Western nations were, in the main, unsuccessful in combating revolutionary warfare. But, beginning in the 1980s, a role reversal became evident. Up to that time, revolutionary warfare was largely the exclusive tool of Marxists. For example, the Soviet Union had supported seventeen different insurgencies in the early 1970s, but no counterinsurgency campaigns. On the other hand, the United States was involved in providing support and assistance to eight different nations that were deeply immersed in counterinsurgency efforts. In the 1980s, substantial U.S. aid to countries managing counterinsurgency campaigns was reduced to only two programs, but American support of insurgent groups involved four separate efforts. At the same time, Soviet advisors, and in some cases troops, found themselves managing or supporting six counterinsurgency campaigns in the 1980s, while their insurgent support list dropped to two programs. Revolutionary warfare had suddenly become a Western specialty.

The American media, sensing a well-organized plot concealed from the U.S. public, ascribed Moscow's Third World troubles to the Reagan administration. Creating the term *Reagan Doctrine,* U.S. broadcasters and print pundits defined their "discovery" as a policy that viewed "enemies of our enemies as our friends." All of this was perfectly suited to TV journalism, a medium that thrives on brash simplicity. It placed complex international events within the familiar context of the Cold War and located the initiator within the White House, a building conveniently containing U.S. media offices. However, the American president had no such simple-minded doctrine.[3]

The phenomenon of revolutionary warfare has been intellectually reduced to a facet of the Cold War at the expense of accuracy and an understanding of more far-reaching implications. The battle between Marxism and the West is only one optional application of revolutionary warfare. Successful prosecution of this form of conflict does not depend on either Marxist or anti-Marxist ideology. It is necessary to attract the fervor of substantial numbers of participants, but they can be motivated by religious beliefs, nationalism, or many other causes. Revolutionary warfare was a natural selection for post–World War II Marxists, but the failure of Marxist economies and the inevitable resistance to Marxist police state methods has brought the era of Marxist revolutionary warfare to its waning days. The 1990s and early years of the 21st Century should bring about a distinct era of Western-supported revolutionary warfare.[4]

Western beliefs, Western economic success, and Western interests all favor a forthcoming age of Western-supported insurgencies in the Third World. Western dependence on diminishing Third World oil supplies and other essential resources increasingly makes free trade an essential Western interest. In the late 1990s and early 21st Century, the West will not long tolerate economically damaging, politically motivated Third World cartels. Nor will the West likely remain passive in an international environment where a major totalitarian power seeks military or diplomatic advantages by manipulating Third World clients.

Western nations are apt to be more assertive in the future southern hemisphere because they will have more advantages there than they enjoyed during most of the 1960s and 1970s. By the early and mid-1990s, it will be obvious in even the most remote regions of the world that actual Western economic achievements far exceed airy promises of those who advocate command economies. In the minds of most of mankind, the battle of economic ideologies will have been settled. Third World leaders who espouse Marxism or other such economic philosophies will have to contend with citizens who increasingly know better. The West will therefore have the ability to make a credible economic appeal to Third World peoples.[5]

That particular Western strength can be coupled with another, a devotion to human rights, including religious toleration. Some ideologies, hostile to Western beliefs, are either intolerant of religion or based on an exclusive religious membership. In either case, the long-term record of Western accommodation of diverse religious views has a practical, popular, and winning appeal. This positive Western belief stands alongside another and growing Western attitude, intolerance of human rights abuses. Based in part on a common notion that restricted freedoms in one part of the world degrade freedom elsewhere, Western states have steadily become more aggressive toward governments that practice political repression. The Western economic sanctions and harsh diplomatic notes of the 1980s are likely to be viewed as too mild for the late 1990s.

These Western trends—a stronger economic and ideological appeal to poorer nations, the growing need for Third World resources, and increasing intolerance of repression—are destined to clash with equally strong trends within the Third World. There, political trends promise more one-party autocratic regimes, more military dictatorships, and a growing use of violence. Third World economic trends portend inefficient extraction industries, environmentally ruinous agricultural practices, and financial management records that discourage capital investment. Lingering Western guilt over the colonial past, and the use of that guilt by Third World apologists, is already wearing thin. To the Western world of the 21st Century, the miscreant dictators of the globe's poorer peoples will be more than merely troublesome; some will likely be intolerable.

But, Western capabilities may not match Western desires. There should be no lack of reasons to foster and support revolutionary warfare in the future Third World. There will be many insurgent leaders who will advocate or, more probably, appear to advocate allegiance to the West and its ideals. There will certainly be no shortage of abusive autocrats in those lands. However, Western governments show little understanding of insurgent movements. They have clearly demonstrated the skill to support guerrillas. Western governments have been able to foster and influence insurgent political leadership groups or underground organizations. Yet, they have failed to develop a mechanism to nurture and support an auxiliary, and give little indication that they even understand the concept. However, the auxiliary, the largest and most important element of an insurgent movement, is an organization that Marxists understand well.

Marxist Revolutionary Warfare

The revolutionary warfare success that Moscow enjoyed during the post–World War II era up to the early 1970s was due to unambiguous strategic policies, a worldwide political party composed of dedicated and often courageous members, a flexible and well-understood tactical doctrine, and a broad base of experience in clandestine activities. The Soviet model for revolutionary warfare is not the only Marxist system. China, Vietnam, and Cuba have all sponsored insurgent movements that differ from some of Moscow's precepts, but their concepts conform to basic principles developed in the Soviet Union. The Soviet model has had its failures, but it has often succeeded and warrants study.

The wellspring of Soviet revolutionary warfare has always been located in a high-level bureaucracy, ensuring Moscow's Third World policy is compatible with Russian interests. Prior to World War II that organ was the Comintern, after the war it was the Cominform, and it is now the International Department of the Communist Party of the Soviet Union, subordinate only to the Politburo.[6] Soviet strategy has often changed in this arena, mainly because Moscow has always reasoned that Third World adventures are secondary to the security of the U.S.S.R., the Soviet economy, and other practical considerations. However, by placing the management and direction of national resources for revolutionary warfare at the highest levels, Soviet leaders have obtained unfiltered reports from and access to the best talents in the field and can support them with adequate diplomacy, funds, materiel, and manpower.

The overthrow of a government is inherently conspiratorial, and the Soviets have the benefit of a wealth of experience in clandestine operations. From its origins, Communism was envisioned as international in application, thereby automatically posing a threat to other forms and philosophies of government. The agents of the international Communist movement were thus ever pitted against national security services and forced into clandestine communications, clandestine recruitment, and sophisticated counterintelligence systems. Through a brutal process of natural selection, only the wily, the courageous, and the dedicated survived. Their legacy has been the methodology for Soviet revolutionary warfare.

The field cadre conducting Soviet revolutionary warfare is composed of Communist party members, a well-disciplined, selfless lot who are international in character and trained to the best standards of their craft. One of their prime and continuing tasks involves recruiting local agents who, in turn, recruit activists. In most cases, the activists do not know they are working in a Moscow-influenced organization, since the standard tactic is to put forward the interests of a "broad front," a revolutionary consortium that absorbs and directs the efforts of the widest possible range of dissident groups. Much of the front's energy is oriented to the creation of the auxiliary at the grass roots and city block level. Initially, the armed guerrilla component is of secondary concern because its members must be spawned from and verified by an established auxiliary. It is this relationship, the indirect link between the party and the auxiliary, and this tactic, the establishment and clandestine control of a front organization, that the West either does not understand, cannot duplicate, or perhaps both.

A successful revolutionary movement—the entire insurgent organization—may comprise little more than 2 percent of a nation's population. Since the armed guerrillas often only account for about 10 percent of the movement, their numbers would amount to roughly 20,000 in a country with ten million inhabitants. In this theoretical example, the auxiliary would number about 160,000 and the underground component might total another 20,000. In order to defeat the guerrilla forces, a beleaguered government of this hypothetical nation might have to establish and maintain a soldier-to-guerrilla ratio of ten or fifteen to one, an army of perhaps a quarter million in strength. These rough numbers are, of course, not applicable to any specific insurgency, but they are representative of a number of actual cases. It is not difficult to see why Marxists settled on revolutionary warfare as a means to extend their influence in the Third World. At the least, a small investment of carefully prepared agents could result in over-extended adversaries and, at the most, a global empire.

The decline in Marxist success in the use of revolutionary warfare was in no way due to failed doctrine. Primarily, the decline came about because Marxism itself came to represent economic and political stagnation. Not only did activists and auxiliary members become more difficult to recruit, Marxists found it increasingly difficult to hold on to the Third World states they gained during the 1960s and 1970s. They became mired in a series of counterinsurgency campaigns and, in some instances, were fighting Western-supported insurgent organizations.

Russian Counterinsurgency, Western-backed Insurgents

Just as Soviet methods of insurgency serve as a good departure point for understanding Marxist revolutionary warfare, Moscow's doctrine for counterinsurgency can be instructive, providing clues to future Western insurgent support policies. While Third World Marxist states are only a small percentage of the authoritarian regimes among the poorer countries, those that exist are well experienced in counterinsurgency. Cuba and the Socialist Republic of Vietnam are daily advising Third World totalitarian states, setting the example in a number of counterinsurgency campaigns. However, those two nations look to Moscow for fundamental doctrine because there is no totalitarian country that has more experience in this field than the Soviet Union.

Paradoxically, the present Kremlin leaders owe much of their counterinsurgency knowledge to their Czarist predecessors. Moscow has been expanding its control of surrounding territory and suppressing revolt for six centuries. There have been a number of setbacks, but the Great Russians have persisted, counted their losses, developed working principles and—in the end—they usually win. The first of the Czarist and Soviet counterinsurgency principles is to place their man in the other fellow's tent. Moscow has long been the master of the penetration agent. The Russians differ from others in that they make massive use of these agents. The second fundamental tenet is to divide and conquer. With or without the assistance of penetration agents, Moscow will bend every effort toward creating schisms between the controlling factions of an insurgent organization. Another principle has to do with the choice of initial concentration. Moscow has always favored going after the urban areas first. The countryside can wait. This step is usually followed by a long-term campaign to win the allegiance of opinion leaders, often by the somewhat medieval practice of taking their young to Moscow for education and indoctrination. Then there is the campaign for the countryside, normally featuring the use of mass deportations. The latter are achieved either by brutal operations that force the peasant from the soil or orderly transport of entire villages to new, tightly controlled settlements.[7]

In most cases, these methods work, but if the costs become too high, the Soviets and their predecessors have always been willing to take a step back to later gain two steps forward. The controlling philosophy in

Moscow has ever been to keep an eye to the long-term goals. The ultimate triumph of world socialism is not doubted, but victory is dependent on the strength of Mother Russia. If supporting either insurgency or counterinsurgency campaigns requires too much, that support is temporarily suspended—but never completely abandoned. Both activities are not seen in the same light as some Americans have viewed them, as "transient and isolated crises."

The Soviets analyze their difficulties in these fields well. One high-ranking military authority probably reached the correct conclusion about Soviet troubles in Afghanistan. Major General Kim M. Tsagolov, a veteran of the fighting there, frankly stated that the Soviet Union had badly underestimated the power of Islam and the Moslem clergy.[8] U.S. assistance efforts to resistance forces in Afghanistan met with considerable success, not because of any particular American strength in supporting insurgency or in the supply of a particular American weapon, but mainly because of ready-made auxiliary organizations, the mullahs and their flocks of faithful. Another factor, directly related to religion, is that there is some evidence that insurgent movements with a heavy religious motivation are capable of avoiding war weariness, an important and needed characteristic for revolutionary organizations engaged in protracted war.[9] It is a characteristic not likely to be readily found among Marxist insurgent groups in the 1990s and early 21st Century.

In some Western quarters, this type of activity, the support of resistance organizations against totalitarian regimes, is an essential 21st Century Western objective. Western involvement in Third World insurgencies that seek to overthrow repressive, autocratic states widens Western strategic options from past choices of defense or retreat to new vistas featuring offensive alternatives.[10] The direction of these activities has heretofore been the special province of the Western intelligence agencies, but during the 1990s the United States changed its approach.

There are two relatively new and American-originated factors that will encourage insurgents and plague counterinsurgency efforts by totalitarian states in the late 1990s and early 21st Century. The first is the National Endowment for Democracy, a private, but government-supported, non-profit organization dedicated to the encouragement of democratic institutions. Founded in 1983, the endowment has supplied newsprint to Third World journals that criticize autocratic governments, funded labor organizations that oppose Marxist states, and printed and distributed books detailing past misdeeds of totalitarian governments. Its growing budgets have broadly based congressional support in the United States.[11] The organization promises to be a goad to rebels and an important thorn in the side of Third World dictators in the mid- and late 1990s.

The second factor that will contribute to the problems of autocratic governments in the Third World is the American innovation of open congressional debate on support of insurgent operations. While the art of overthrowing troublesome governments is a time-honored feature of statecraft, voluntary public disclosure has been taboo—until the 1980s. Growth in the power of the U.S. Congress in that decade transformed support of insurgents from a covert and almost exclusively executive branch function to an open activity of shared responsibility between the Office of the President and the legislative branch. Ostensibly, the change was aimed at tempering ill-advised administration adventures in the Third World. The actual effect has been something quite different. Prospective insurgent leaders found their pleas no longer limited to discrete U.S. government officials. Instead they could offer their schemes to a wide array of different-minded legislators as well. Traditional congressional tactics such as "log rolling" came into play, one group of legislators offering their votes for an insurgency support proposal on the condition that their pet guerrilla assistance project receive the votes of another group. Then, too, the whole process came out of the darker corners of Washington so that faulty procedures could be identified, exposed, and corrected in a way that produced the best chance of securing approval.

Western-supported insurgencies, particularly those backed by the United States, are directed at Third World totalitarian states that increasingly use Russian-based counterinsurgency doctrine. These conflicts do not necessarily have their roots in the East-West ideological confrontation or in superpower geopolitics. They may initially represent little more than localized civil wars. But, in an ever more interconnected and interdependent world, events in some faraway place that would have seemed obscure and irrelevant in the 20th Century are likely to attract attention and action in the 21st Century. Western-sponsored insurgencies against Third World states using Soviet counterinsurgency techniques will probably be a pronounced feature of the late 1990s and early 21st Century.

Western Counterinsurgency Methods

While Western nations, particularly the United States, are increasingly involved in the business of bringing down dictatorial governments in the Third World, there are still some Western-supported counterinsurgency campaigns. Despite their decline in numbers, these efforts are important to the West because the successful conduct of, or assistance to, a counterinsurgency campaign will determine a degree of Western influence. The West has prosecuted these efforts in a markedly different way than Marxist states, and the overall Western record is not inspiring.

One would think the steady decline in Western-conducted or supported counterinsurgency campaigns constitutes a favorable trend for North America and Western Europe. After all, by the late 1980s the Soviet Union was maintaining thirty times the number of military advisors and trainers in the Third World as the United States fielded. In dollar terms, Soviet military aid to Third World states was five times U.S.

figures in the same category. But, Third World military students in the Soviet Union are twice the comparable figures in the United States, and there are concerns over a rapid decline in Western military influence among the poorer nations.[12] Whatever the case, it is just as well that the West is no longer deeply involved in counterinsurgency campaigns because there are systemic obstacles to effective Western support or conduct of such operations.

Effective prosecution of a counterinsurgency campaign requires the use of methods that are antithetical to core Western beliefs. Since war is usually not declared in these cases, the rule of law and normal bureaucratic relationships must be observed. Western soldiers involved in counterinsurgency have grown accustomed to finding themselves at war without essential wartime authority.[13] Fundamental counterinsurgency programs aimed at undercutting the appeal of the insurgent, such as land reform, propaganda, and civic action efforts, may have the active involvement of as many as four or five different bureaucracies, each insisting on its own institutional imperatives. Armies that become immersed in village level politics, agricultural programs, and police functions have a great tendency to become politicized, an anathema to Western notions of military propriety.[14] Additionally, the basic elements of counterinsurgency campaigns must be carried out concurrently, not in the sequential fashion that Western officers use to conduct conventional operations. Police, civil reconstruction, and economic development tasks all have to be accomplished simultaneously and coordinated with military operations.[15] All told, counterinsurgency does not rest easy with Western military thinking or Western norms.

These factors and more are impediments to the successful accomplishment of a counterinsurgency effort, and there is no trend indicating the West will be any more adept at this type of activity in the 21st Century than it was in the 20th. In fact, there are indications that, at least for the United States, assistance to Third World states involved in counterinsurgency campaigns will be a more difficult enterprise. Two U.S. laws, passed in 1988, are bound to hamper American efforts to aid nations battling insurgents. The first would penalize any nation that has taken measures to repress industrial strikes. The fomenting of workers' strikes is a central activity on any insurgent's urban agenda, and strike abatement is normal in a besieged government's counterinsurgency effort. The other law requires foreign governments to accede to American inspections of any and all records and files relating to a U.S. arms transfer, regardless of foreign domestic laws, security interests, or sensitivities.

These laws were added to other post-Vietnam legal restrictions such as the War Powers Act. While American presidents have questioned the constitutionality of that act, they have usually complied with it. It restricts an American assistance effort by prohibiting members of the U.S. armed forces from intentionally entering combat areas without a presidential notice being given to the Congress. In effect, it prevents American military observation of an aid recipient's combat performance, drastically reducing the knowledge necessary to conduct tailored, realistic training and assistance programs. Taken together with other American restrictions on aiding foreign police and internal security services, drug-related legislation, and various U.S. human rights demands, counterinsurgency is now and probably will continue to be an extremely constrained U.S. military activity in the late 1990s and early 21st Century.[16]

Among Western nations, only Britain has proven to have a more or less consistent record of military success in Third World counterinsurgency campaigns. But, the United Kingdom's efforts in this field will probably be limited, as they have been in the past, to its former colonial possessions. American military assistance performance in counterinsurgency programs may actually have deteriorated. Although the United States had a number of counterinsurgency assistance successes in Latin America during the 1960s, its experience in that region all throughout the 1980s has been comparatively poor. Seeking reasons for the lack of success, several American officers produced an analysis that cited the lack of attention to counterinsurgency in the U.S. military educational system, a lack of centralized responsibility in Washington, a cumbersome and unresponsive military assistance system, the use of inappropriate technology, and a proclivity on the part of the armed services to assign inferior talent to Third World counterinsurgency programs.[17]

There are few American developments designed to improve performance in this type of activity. It is, perhaps, revealing to note that some of those that have been forwarded concentrate on minimizing the exposure of American military personnel. For example, the ARAPAHO concept, the use of off-shore container ships configured to handle helicopters, is envisioned as providing a secure and distant base for American military assistance organizations, thereby avoiding the use of highly visible and vulnerable land bases. Another concept is called *echelonment*, the forward deployment of only a small part of a military organization in a low-intensity conflict zone. A unit would keep the great majority of its headquarters, administrative, and organizational support contingents at its home base, accomplishing internal communications by use of dedicated satellite radio links. This latter concept has been tested in Latin America and is believed to hold even greater promise with the use of a forthcoming technology, interactive image communications.[18] Both concepts have the goal of putting as much distance as possible between American forces and the area of conflict, hardly an orientation to victory. But such thinking evidently has the support of important members of the U.S. Senate, some of whom have called for "more effective means of indirect involvement to protect U.S. interests and promote democracy in the Third World."[19]

There are few Western material innovations that are actively being pursued in the field of military counterinsurgency. An exception is an air-activated sound-detection device, designed to be employed in eavesdropping. Using fluidic and acoustic technologies, the passive device can differentiate between human voices and other sounds. It could possibly be employed in areas of known guerrilla concentrations.[20] In some ways, this program is merely attempting to regain some of the technology that was readily available during the Vietnam War.

Insurgency and Counterinsurgency in 2010

The decline in Marxist political and economic fortunes may foretell an end to the era of global ideological competition, but there is small hope for an end to Third World insurgency and counterinsurgency. Another era will begin, and trends clearly indicate more—not less—interaction between nations and blocs of nations. Not all of these increased relations will be friendly, or even peaceful. Although there is an ongoing reduction in the incidence of major mid-intensity conflicts, there is an undeniable rise in the incidence of low-intensity conflicts, most of them armed insurgencies. In a world that has ever-expanding interdependence and growing connections between diverse regions, it is likely that the major powers will be involved in both the insurgencies and the counterinsurgency efforts.

In the world of 2010, Third World insurgents, particularly those supported by the West, will probably have the upper hand. Certainly the technology—shoulder-fired anti-aircraft and anti-tank weapons, improved clandestine communications equipment, clothing, footwear, body armor, and lightweight rations that are vastly superior to currently available items—and RPV resupply systems favor the insurgent and weigh against the counterinsurgent. Added to these material aids, the 21st Century insurgent may be able to employ unattended anti-helicopter mines, currently under development in the United States.[21]

However, heavy-handed Western assistance can be counterproductive in Third World insurgency support situations. The downing of an American-piloted air drop plane in Nicaragua in the mid-1980s severely damaged U.S. policy there, because there was an over-reliance on technological solutions for logistical problems. At about the same time, in Afghanistan, American mules were being used to support resistance groups. Twenty mules are capable of moving 5,000 pounds a day for fifteen miles in difficult terrain.[22] The more sophisticated approach to solving insurgent problems does not always produce the best results. Altogether, the Western-supported insurgent has an advantage—if the right Westerners are chosen for the support roles.

There is also a growing political benefit to the Western-oriented insurgent. Western nations, insatiable in their need for raw materials and more and more insistent about international human rights standards, will probably take an even greater interest in southern hemisphere nations. The West may be more inclined to support efforts to overthrow an abusive Third World dictator in 2010 than it is today. Systemic obstacles to supporting or using force abroad are largely contingent on the nature and conduct of a recipient country. At least initially, an insurgent group cannot be held accountable for the conditions within a country. Its leaders will promise better conditions. The insurgent organization naturally becomes a more worthy Western client than a nation conducting a counterinsurgency effort.

Third World insurgents in the 21st Century would do well to study Soviet methods of insurgency, particularly the techniques to recruit and employ an auxiliary. It appears that the Soviets will not have much need for their insurgency doctrine in the late 1990s and early 21st Century, but the fundamentals are sound and there is no reason basic Soviet insurgency doctrine should not work in 2010. One modification, an enlargement of the underground or, alternatively, more concentration on an urban auxiliary, seems warranted. Rapid urbanization of the Third World makes this change necessary.

The heyday of Soviet-inspired insurgency did not see the massive, Third World urbanization so prevalent now, a characteristic providing a fertile ground for insurgency. Despite oversaturation in some of these cities, more is promised. The Middle East is expected to be 73 percent urban in 2010 and a city such as Cairo, now having a population of eight million, is expected to reach twenty million inhabitants in that year. Much of this growth is attributed to migration patterns of unemployed male youth, a sector of the population that can easily be aggressive and belligerent. During the 1970s and 1980s, there was some relief for Third World governments due to immigration policies in both Western Europe and the United States. However, that "safety valve" began to close in the late 1980s, European and American governments tightening their entry requirements. Then, too, growing Third World unemployment and underemployment (40 percent in Latin America) foreshadows growing discontent in the poorer cities of the globe.[23]

Although there will be many future insurgent leaders who will look to the West for support, there appears to be no reason for these prospective leaders to expect any aid in organizational concepts. Western states may be adept at supplying the guerrilla component after it is raised, but Western governments do not seem to grasp revolutionary warfare and have no mechanism to create or stimulate an insurgency in the way the Soviets have time and again demonstrated.

Soviet doctrine in opposing the insurgent is also likely to be useful in 2010. The West has a flawed past in counterinsurgency campaigns, and its history of assistance to Third World states combating insurgents is equally weak. Additionally, Moscow is the logical patron for Third World totalitarian governments, regimes that are most prone to experience armed revolt. Completely apart from the issue of ideology, the Soviet

Union's long record of successfully suppressing insurgencies has much to offer autocratic states facing internal rebellion.

However, some Soviet counterinsurgency methods are not usable by the West, either in assistance situations or in the case of direct use of Western military forces. For example, mass deportations of Third World populations are unlikely to be tolerated by Western peoples. The U.S. armed forces in particular are constrained in the field of counterinsurgency assistance. The prohibition on aiding police or internal security services, the War Powers Act that restricts first-hand American military knowledge of indigenous combat performance, and other laws severely hamstring U.S. aid and assistance to a nation fighting guerrillas. If the West is to have an effective counterinsurgency capability in 2010, it must seek fresh approaches to the problem.

There are two central causes of the generally poor Western military record in the field of counterinsurgency. The first is that Western armies are either not large enough or do not consider it important enough to maintain a full-time, well-qualified cadre for counterinsurgency tasks. This is perhaps a good choice, because the main task for these organizations is to ensure an adequate response in the event of higher forms of conflict. The resulting cost, of course, is to occasionally field partially qualified novices in counterinsurgency situations where professionals are required. The second cause of lackluster Western military performance is that Western peoples will not long tolerate the use of their soldiers in suppressing rebellions in a distant land, whether their soldiers are in a direct combat role or serving as advisors. Unfortunately, if the past is any guide, Western nations will probably find it in their interest to participate in some counterinsurgency campaigns. The solution for these problems may lie in taking the military aspects of counterinsurgency campaigns out of the hands of Western military forces.

An international corporation composed of former Western officers and soldiers skilled in acceptable counterinsurgency techniques would largely solve both of these Western counterinsurgency problems. Western aid to a country conducting such campaigns could be restricted to yearly grants or loans, contingent on two conditions. The recipient government would have to contract for the corporation's services, and it would have to demonstrate a good faith effort to comply with the corporation's advice and recommendations. In keeping with the Western tenet that, in the end, the recipient has to win its own war and cannot have someone else win it, that nation should decide for itself how much of the aid is spent on economic development programs and how much is devoted to military forces and military operations. The role of the corporation would involve expert advice, training, and assistance, not combat. Considering the record of most Western governments in the field of counterinsurgency, the corporation would not have to work very hard to achieve comparatively superior results. And a commercial concern would likely attain those improvements at considerably less cost.

Decisions for Western leaders would be a bit easier to make. Too often the West has thrown good money after bad because Western military personnel were involved and thus Western prestige was at stake. The major benefit, however, would be that Western military forces could free themselves of this onerous burden, one they have not borne well.

In the pentapolar world of 2010, there will be more interdependence and thus more relations among nations. The major powers will attempt to manipulate events within the Third World, and there are those in the Third World that will attempt to play on the needs of the major powers. These North-South relations will on occasion feature major power support of clandestine insurgency and counterinsurgency campaigns. Trends indicate many of the insurgency efforts will be Western-supported, pitted against Russian-supported counterinsurgency forces. Thus, these events will retain a characteristic of 20th Century conflicts, East-West competition. Currently, neither the East nor the West appears to be well prepared for these conflicts. Success or failure is apt to turn on the adjustments in policy and capabilities that must be made in a changing environment between now and the early years of the 21st Century. Those adjustments will have much to do with the type of world the children of the 20th Century inherit for their lives in the 21st Century. However, there are other forms of low-intensity conflict that will shape the world of 2010.

Notes

[1] Fred C. Iklé et al., *Discriminate Deterrence: Report of the Commission on Integrated Long-Term Strategy* (Washington, D.C.: U.S. Department of Defense, 1988), p. 15.

[2] U.S. Army and U.S. Air Force Low Intensity Conflict Study Group, *Joint Low Intensity Conflict Project Final Report* (Ft. Monroe, VA: Training and Doctrine Command, 1986), pp. 8–9.

[3] Ambassador Jeane Kirkpatrick's "Protracted Warfare: The Third World Arena" presentation at the 16th Annual International Security Studies Program, Boston, Massachusetts, April 22, 1987. Also, see Assistant Secretary of Defense Richard Armitage, "Anti-Communist Insurgencies," *The Retired Officer* (January 1986): 21.

[4] Steven D. Goose, "Low Intensity Warfare," in Michael T. Klare and Peter Kornbush, eds., *Low Intensity Warfare: Counterinsurgency, Proinsurgency, and Antiterrorism in the Eighties* (New York: Pantheon Books, 1987), pp. 80–111.

[5] Office of the President, *National Security Strategy of the United States* (Washington, D.C.: The White House, 1987), p. 34.

[6] Richard H. Schultz, Jr., *The Soviet Union and Revolutionary Warfare: Principles, Practices, and Regional Comparisons* (Stanford, CA: Hoover Institution Press, 1988), pp. 16–31.

[7] Rod Paschall, "Marxist Counterinsurgencies," *Parameters* (Summer 1986): 5–6.

[8] Major General Kim M. Tsagolov, as quoted in the *New York Times* (July 24, 1988), p. 11.

[9] Farid Abolfati, "Assessing War Weariness in Insurgencies" (Carlisle Barracks, PA: U.S. Army Military History Institute [USAMHI], 1987), p. S-3.

[10] Paul F. Gorman et al., *Supporting U.S. Strategy for Third World Conflict* (Washington, D.C.: U.S. Department of Defense, 1988), p. 19.

[11] Carl Gershman, President, National Endowment for Democracy, as quoted in Joseph Pichirallo, "How the U.S. Supports Nicaraguan Opposition," *Washington Post* (September 29, 1988), p. A19.

[12] Paul F. Gorman et al., *Commitment to Freedom* (Washington, D.C.: U.S. Department of Defense, 1988), p. 10.

[13] U.S. Army Center for Land Warfare, "Theater Planning and Operations for Low Intensity Conflict Environments" (Carlisle Barracks, PA: U.S. Army War College, 1986), p. i.

[14] Ian F.W. Beckett and John Pimlott, eds., *Armed Forces and Modern Counterinsurgency* (New York: St. Martin's Press, 1985), pp. 10–13.

[15] Sealon R. Wentzel, "Operational Art and Low Intensity Conflict" (Carlisle Barracks, PA: U.S. Army War College, 1987). For a sound analysis of Western imperatives in a counterinsurgency support effort, see Peter Bahnsen and Captain William H. Burgess III, *U.S. Aid to Democratic States Facing Totalitarian Revolutionary Warfare: Twelve Rules* (Langley Air Force Base, VA: Army–Air Force Center for Low Intensity Conflict, 1987), p. v.

[16] General Fred F. Woerner, Commander-in-Chief, U.S. Southern Command, "The Strategic Imperatives for the United States in Latin America," *Military Review* (February 1989): 18–28.

[17] Lieutenant Colonel A. J. Bacevich et al., *American Military Policy in Small Wars: The Case of El Salvador* (McLean, VA: Pergamon-Brassey's, 1988), pp. 49–51.

[18] Gorman, *Supporting U.S. Strategy for Third World Conflict,* pp. 44–50.

[19] Letter of Senators Sam Nunn, John W. Warner, Edward M. Kennedy, and William S. Cohen dated January 25, 1989, to Lieutenant General Brent Scowcroft (U.S.A.F., Ret.), Assistant to the President for National Security Affairs, as reported in *Armed Forces Journal International* (March 1989): 66–67.

[20] "Undetectable Listening Device for U.S. Army," *Jane's Defence Weekly* (January 28, 1989): 116.

[21] "Helicopter Mine Project to Start," *Jane's Defence Weekly* (January 28, 1989): 118.

[22] Field Manual 25-7, *Pack Transportation* (Washington, D.C.: Department of Army, 1952), paragraph 3C.

[23] Hernando De Soto, *The Other Path: The Invisible Revolution in the Third World* (New York: Harper and Row, 1989), p. 234. Andrew W. Marshall et al., *The Future Security Environment* (Washington, D.C.: U.S. Department of Defense, 1988), pp. 14–15.

CHAPTER EIGHT

Low-Intensity Conflict: Peacemaking and Peacekeeping

Not long before his tragic death, Dag Hammarskjöld, the late secretary general of the United Nations, reflected on the role of the world body in peacekeeping. He said he saw that role as ". . . an international version of containment designed to restrict the expansion of the zone permeated by bloc conflicts."[1] Unfortunately, the era of Hammarskjöld's service was one marked by the failure of the United Nations to achieve peaceful resolution of a number of conflicts. The zone permeated by bloc conflicts was global in scope because the Western and Eastern blocs viewed conflict as a zero-sum game. A gain for one side (the prize usually being some Third World state) was believed to be a loss for the other contender. However, when the major powers did agree on a peaceful resolution of some armed dispute, they had the power and influence to end the violence. The Cold War is now ending, and one could project an end to ideological competition, diminished control exercised by superpowers, and less enthusiasm for peacekeeping.

But, international efforts to secure peaceful stability are growing because conflict is not necessarily a zero-sum game. It can easily be a negative-sum game in which the number of losers exceeds the number of winners. Wars disrupt the normal flow of trade, interrupt debt payments, destabilize markets, and contribute to the spread of disease. The condition of armed conflict encourages leaders to suppress dissent and ignore internationally recognized standards of individual rights. War and famine are often linked. Wars are contagious. They spill over into otherwise neutral states, drawing in unwilling participants. As such, wars can and often do degrade the economic and physical well-being of the world community. A global nuclear war could, in fact, destroy that community. It is therefore in the interest of mankind to curtail, arrest, limit, or stop armed conflicts. Peacekeeping and peacemaking are rapidly expanding activities in the late 20th Century, and both are likely to have a healthy future in the early 21st Century. Hammarskjöld's notion of "containment" is likely to grow—with or without the existence of bloc conflicts.

Stability Operations: Peacemaking and Peacekeeping

The term *peacekeeping* is often applied to any externally imposed military activity aimed at peacefully resolving armed conflict. For example, the failed British, French, Italian, and U.S. military effort to bring an end to violence in the streets of Beirut, Lebanon, during 1983 was mislabeled "peacekeeping." There was fighting before, during, and after the allied deployment of troops. In fact, there was no peace to keep. A more accurate description of the allied effort there would have been *peacemaking*. These Western nations attempted to bring about peace by using force.

Peacekeeping is an activity designed to maintain a state of peace.[2] It is meant to preserve the absence of war. Peacekeeping does not necessarily involve military forces; the term can be applied, for instance, to diplomatic endeavors. Peacemaking, on the other hand, assumes violence is in progress. The term *peacemaking* may also characterize diplomatic efforts, but it can describe the use of violence if that violence is externally imposed and aimed at ending an ongoing conflict. Both terms, *peacekeeping* and *peacemaking*, when used to identify military activity, can be grouped under a more all-encompassing term, *stability operations*.

Neither peacekeepers nor peacemakers have to be moved by altruistic motives. Participation in a stability operation may stem wholly from narrow, selfish interests. Many nations desiring reliable, uninterrupted supplies of oil from the Middle East, for example, are willing to support both peacemaking and peacekeeping in that troubled region. The economic stakes there are indeed high. An external force, employed in stability operations, can therefore be a "winner." The United States "won" two peacemaking efforts in the Caribbean region when it used force to end violence in the Dominican Republic in 1965 and again on the small island of Grenada in 1983. Conversely, the Soviet Union "lost" in its 1979 effort to rescue and reform its besieged client regime in Afghanistan, possibly attempting to repeat the success Moscow had enjoyed in Czechoslovakia during 1968. Peace, secured at the right moment, can secure national and international advantages.

Absolutely effective peacekeeping operations should not be grouped under the category of low-intensity conflict, because no conflict will have occurred. However, since this condition is so rare, it is logical to anticipate some violence and include peacekeeping as a category of low-intensity conflict. Peacemaking that features combat between regular armed forces falls outside the category of low-intensity conflict for obvious reasons. This latter condition is also rare, since few nations are willing to commit their forces in the midst of an ongoing mid-intensity conflict to bring an end to the fighting. Stability operations are therefore logically placed in the general category of low-intensity conflict.

Most peacekeeping operations in the late 1990s and early 21st Century are likely to continue the pattern established in the late 20th Century, efforts conducted under the auspices of the United Nations. Almost all of the fourteen U.N. military adventures during the organization's first forty years were peacekeeping tasks, and most were observer missions. Indeed, the U.N. administrative and schooling apparatus for peacekeeping is based on the assumptions that a truce has been reached between two or more belligerents, a neutral zone has been agreed to by those belligerents, and the U.N.'s business is little more than to report infractions of the cease-fire. The U.N. Peacekeeping Manual and the yearly training course at the U.N. Peace Academy are both structured along these lines.[3]

Many stability operations have been conducted outside the purview of the United Nations, and this will probably continue to be the case well into the future. Major powers, acting in accordance with their own interests, avoid anticipated opposition or delay in the U.N. Security Council or General Assembly and opt for unilateral peacemaking operations aimed at securing a truce or ending an armed dispute by the use of force. Examples of this type of action include the British use of force in Gambia during 1982, brilliantly led by Lieutenant Colonel Ian Crooke, and the Indian government's deployment of troops to Sri Lanka in 1988. Additionally, major powers may also continue the practice of organizing a regional coalition of military elements to conduct stability operations. The U.S. sponsorship of eastern Caribbean military forces to conduct peacekeeping on the island of Grenada after the American intervention there during 1983 serves as an example of this type of stability operation.

Growth of 21st Century Stability Operations

Peacemaking and peacekeeping can be expected to increase during the late 1990s and early 21st Century. The anticipated rise in the incidence of stability operations is in part based on the demise of the bipolar world and in part on the need for stability in regions rich in raw materials. It is in some measure due to the growing North-South debt and investment links, but also partially due to increased international concerns about human rights abuses under repressive governments. The expansion of stability operations is, in some respects, a natural result of an ever more interdependent world. But there is another, and less obvious, reason to expect future growth in the numbers of stability operations. The international community applauds the peacemaker, and there have been national leaders in the recent past who have taken domestic political risks to settle armed disputes. With praiseworthy 20th Century examples, one can expect repeat performances in the 21st Century.

Instability, a prime feature of the East-West confrontation of the mid- and late 20th Century, is not likely to be welcomed during the early 21st Century. In the earlier period, the spread of Marxism, a politico-economic system that must be imposed by force, was dependent on the existence or creation of instability. Moscow had a vested interest in fostering and maintaining unrest in target nations. In opposition, the West found itself defending the *status quo*. Since the Soviet Union often used its position on the U.N. Security Council to block conflict resolution actions, development of the planned post–World War II international security system was severely hampered. The solution became bloc alignment. Each superpower raced the other in gathering client states. The end result was that each side found itself supporting troublesome, dependent, and demanding allies. Perhaps reassessing its economic and military liabilities, the Soviet Union began supporting Third World peacekeeping initiatives in the late 1980s and finally proposed a sweeping concept for an international apparatus designed to end the unwritten ban on major power peacekeeping contingents. The U.S.S.R. also advocated a stand-by U.N.-controlled peacekeeping force.[4] Essentially, Moscow made a move toward global *status quo* and the use of the U.N. as it was originally envisioned in 1945.

U.N. peacekeeping efforts could easily be more effective in the 21st Century than they have been in the 20th, and thus more likely to be employed. With probable cooperation between the Soviet Union and the United States, effective use of the Security Council's Military Staff Committee can be expected. This moribund body, identified in the U.N. Charter, was to have been composed of the military chiefs of staff of the permanent members of the Security Council. The committee, meeting regularly during the contentious Cold War period, never had a substantive issue on its agenda.[5] But, the committee could be a useful organization in speeding planned and coordinated deployments of peacekeeping forces.[6] Heretofore, military leaders have only been able to plan and react to decisions after lengthy debates in the General Assembly or extensive negotiations by the secretary general's office. The result has been that essential military questions have not been addressed until the very eve of employment. This has caused delay, confusion, inappropriate equipment choices, and questionable performance. An active Military Staff Committee would be able to produce plans, politico-military options, and ready forces concurrent with political negotiations.

Twenty-first-Century North-South relations are apt to produce a greater use of stability operations. The permanent members of the U.N. Security Council are all northern hemisphere nations, while the vast majority of conflicts occur in the southern hemisphere and lower tier of

an observer. The technology of 2010 can significantly assist such a peacekeeper.

The Limits of Stability Operations

There are limits to the effectiveness of both peacekeeping and peacemaking operations, and most of these limits will probably be carried forward into the 21st Century. Among the limits are costs, lengths of employment, degradation of force effectiveness, and geographic restrictions. There is also no indication that future stability operations will be any more effective in suppressing some forms of violence in the 21st Century than they were in the 20th. Additionally, peacemaking operations will likely be limited by restrictive overflight and intermediate base use conditions. Peacekeepers operating under the auspices of the United Nations will still have to contend with the inflexible and limiting U.N. command and control arrangement. But, there is some hope for constructive solutions.

Peacekeeping can be expensive, particularly when national armed forces are the designated peacekeepers. A single guard post, manned around the clock, costs the United States Army between $80,000 and $120,000 per year.[15] These expenses include the necessity for guard reliefs, a share of supervision expenses, portions of the guards' pay, medical and retirement costs, and training overhead. There are other costs associated with use of the world's armed forces as peacekeepers. These organizations are formed for purposes other than peacekeeping, and those original purposes are not served while a nation's military units are deployed and engaged in peacekeeping tasks. That nation's capability to wage or deter mid- or high-intensity conflict, for instance, is diminished. Then, too, peacekeeping is a mind-numbing, boring duty that degrades a combat unit's fighting edge.[16] Obviously, there are more cost-effective solutions to peacekeeping tasks than employing national armed forces, yet the standard response to the call for peacekeeping has been to use these organizations.

National armed forces often have to be retrained for peacekeeping tasks. The U.N. Training Center staff at Niinisalo, Finland, has found it necessary to teach area studies, transportation, communications, equipment identification, mine identification and removal, medical procedures, common terminology, investigation, and conflict resolution subjects. The staff is constantly plagued with language barriers, as the armed forces of fifty-plus nations have been employed in peacekeeping tasks at one time or another and there is no predictable or even rational selection process for peacekeeping forces.[17]

Peacekeeping can be expensive, in terms of both immediate costs associated with a given operation and expenses over time. If the peacekeeper is expected to intercept violators of an air, land, or coastal boundary, peacekeeping costs can be prohibitive. The Israeli border barriers, tank ditches, electronic fence, and balloon-borne electronic surveillance systems are effective but costly. A similar but less comprehensive barrier constructed for drug interception along the southern limits of the United States during the 1980s has also been costly, but not nearly so effective. It has been estimated that the U.S. southern intercept zone only interrupts 5 to 7 percent of illegal drug traffic.[18] These costs are high not only in terms of deployed force structure and barrier investments, but also in terms of time. Some peacekeeping elements in the Middle East have been in operation for twenty years or more.

Having to pay 57 percent of the U.N.'s peacekeeping bills, the five permanent members of the Security Council formally protested in 1988 when they learned costs were going from $300 million per year to $2 billion in the space of twelve months.[19] Some relief may be in sight because Japan, a nation that has flourished under a largely foreign-financed security umbrella, has demonstrated a commendable willingness to make a peacekeeping financial contribution in addition to its assessed U.N. dues.

United Nations–sponsored peacekeeping operations have particular limits. A major power such as the United States is not likely to welcome a peacekeeping scheme of the U.N. General Assembly or even the secretary general's office in the Caribbean or Central American regions. The same can be said of a Soviet reaction to such plans for Eastern Europe. Both nations are likely to preempt U.N. plans by unilateral action or by use of the Organization of American States or the Warsaw Pact structure. Additionally, U.N. peacekeeping actions are marked by a rigid and highly bureaucratic control system that prohibits peacekeeping force commanders from negotiating substantive disputes at the scene.

Both peacekeeping and peacemaking operations share a common limitation, lack of effectiveness against a substantial insurgent organization. External forces have an exceptionally poor record against rural insurgents, and they do not have the local expertise to root out a healthy urban underground movement. Peacemakers and peacekeepers do best against uniformed opponents or weak, ill-organized resistance forces.

Major-power peacemaking endeavors have a limitation not normally experienced by peacekeeping forces, difficulties with base and overflight rights. Depending on the location of the scene of instability, intermediate refueling points and overflight clearances may be essential for deployment of a force directed to bring about an imposed peace. Since these operations are often initiated by a northern developed nation seeking a favorable outcome at some spot in the south, regional powers are normally a bit wary of externally imposed solutions, as they fear they themselves may be the next target.[20]

While peacemaking operations require the use of a quickly deployed national armed force, operating to achieve national objectives, peacekeeping forces are not in the same category, and therein lies a partial solution to 21st Century peacekeeping problems. There is no reason that peacekeeping forces cannot be supplied, trained, and employed by a commercial organization. Nations that support peacekeeping operations would not lose the services of a part of their armed forces, and national military elements would not have to be retrained and restructured for peacekeeping tasks. Commercial peacekeeping forces could certainly bring a number of efficiencies. These elements would not have to bear the large overhead associated with military organizations that have to field combat forces, units that must cope with mid- and high-intensity conflict. They could produce observation teams that are tailored at their inception for a particular peacekeeping task. The problems for the U.N. Training Center could be considerably reduced. And a professional peacekeeping unit's performance would probably be superior to that of an organization that understandably views peacekeeping as a secondary function. The advantages of commercial peacekeeping forces are strong enough that their emergence sometime in the 21st Century is probable.

Stability Operations in 2010

The stability operation is little more than a tool of statecraft. A commitment to peacekeeping by a country securing a truce at an opportune time is usually advantageous to that nation. If enough nations are benefited by a peacekeeping operation, there may be utility in placing the operation under the direction of the United Nations. Conversely, a proposed supervised truce may jeopardize some other nation's position so that it will not respond to a call for participation in peacekeeping activities. It is all a matter of national interests. Dag Hammarskjöld's vision of "international containment" is not likely to become a sweeping movement of the early 21st Century. Nevertheless, a reduction in ideological confrontation is likely to result in the growth of peacekeeping operations.

Hopefully, the costs of U.N. peacekeeping operations will not be a seriously limiting factor in the 1990s and early 21st Century. The Japanese government should be encouraged to increase its peacekeeping funding and to accept a share of the U.N.'s peacekeeping management burdens. Largely due to the constant encouragement of the United States, Japan broke the 1 percent of GNP barrier for its military forces in the late 1980s. Yet no state, including Japan itself, seeks a rearmed Japanese nation. The steady growth of the Japanese economy shows every prospect of continuing, so a logical policy for Tokyo might involve the support of peacekeeping as opposed to a further growth of Japan's armed forces. That policy would clearly match the desires of most of Japan's people and promotes a better international image for Tokyo.

Peacemaking operations, largely executed by developed nations in underdeveloped regions, will probably grow. Rooted in national interests, the desired outcomes of these enterprises can often bring about what most nations seek—security, influence, favorable trade, and power.

There are any number of reasons to expect a larger number of peacekeeping operations in the future. Concern over access to resources and the elimination of human rights abuses will probably figure heavily in such 21st Century North-South deployments. Effectiveness is in some measure dependent on the quick application of overwhelming force. While a strong amphibious capability would be useful, a large-volume military air transport fleet is of more utility, and the Soviet Union appears to be outpacing the West in that field.

It is important to do peacekeeping well, and there is much room for improvement. United Nations–sponsored peacekeeping functions could be enhanced by using the Military Staff Committee of the Security Council as it was originally intended to be used. Peacekeepers, whether they are operating under the auspices of the U.N. or not, need more freedom of action in the field. They need direct, immediate communications links with the government of potential belligerent A, for example, so that menacing movements of government B's forces can be expeditiously reported, and vice versa. Peacekeepers need the apparatus and technology of the Israeli border security system. They also need investigative techniques, so that potential violators of a truce face the threat of being exposed. Above all, peacekeeping must become a well-developed skill instead of a part-time secondary duty. Leaving such a vital task to the whims of some nation's military forces is a certain path to subpar performance. Peacekeeping is a function that would be better done by a commercial organization composed of professionals.

Notes

[1] Inis L. Claude, *Swords into Plowshares* (New York: Random House, 1972), p. 224.

[2] Another, more restrictive, definition is found in Major General Indar Jit Rikhye's *The Theory and Practice of Peacekeeping* (New York: St. Martin's Press, 1984), pp. 1–2.

[3] Major General Indar Jit Rikhye, Director, U.N. Peace Academy, as quoted in the *New York Times,* 1988, July 19, p. A4.

[4] Paul Lewis, "Soviets Urge Nations to Provide a U.N. Army," *New York Times,* October 3, 1988, p. A4.

[5] Rikhye, *Theory and Practice,* p. 3.

[6] Former President of the U.N. Security Council Ole Algard, *New York Times,* July 19, 1988, p. A4.

[7] Dimitri K. Simes, Carnegie Endowment for International Peace, "If The Cold War Is Over, Then What?" *New York Times,* December 27, 1988, p. A21.

[8] Major David T. Zabecki, "The Congo Operation: A Case Study in the U.N. Peacekeeping Model," unpublished monograph (Carlisle Barracks, PA: U.S. Army War College, 1988), pp. 1–5.

[9] John R. Allen, "Peacekeeping and Local Presence Missions: Capabilities and Challenges," *Defense Science 2003* (December/January 1986): 54–62.

[10] Frank C. Carlucci, *Soviet Military Power* (Washington, D.C.: Department of Defense, 1988), pp. 93, 134.

[11] *Jane's Defence Weekly* (December 10, 1988): 1439.

[12] Spokesman, National Aeronautics and Space Administration, as quoted in the *New York Times,* May 18, 1988, p. B9.

[13] General Michael Dugan, U.S. Air Force, as quoted by Barbara Amouyal, "Air Force to Stretch C-17 Production to Cut Budget," *Defense News* (March 6, 1989): 3.

[14] Author's personal observations in Israel in 1986.

[15] U.S. Army, *Research and Development Plan for Army Applications of Artificial Intelligence Robotics* (Menlo Park, CA: SRI International, 1983), pp. 1–5.

[16] James G. Hunt and John D. Blair, eds., *Leadership on the Future Battlefield* (Washington, D.C.: Pergamon-Brassey's, 1985), pp. 208–209.

[17] Marrack Goulding, U.N. Under Secretary General for Political Affairs, as quoted in *Jane's Defence Weekly* (December 24, 1988): 1589.

[18] Commandant of the U.S. Coast Guard, Admiral Paul A. Yost, Jr., as quoted in the *New York Times* (December 8, 1988), p. A23.

[19] *New York Times* (December 28, 1988), p. A5.

[20] Major Bradley L. Butler, "Planning Considerations for the Combat Employment of Air Power in Peacetime Contingency Operations" (Langley Air Force Base, VA: Army–Air Force Center for Low-Intensity Conflict, 1988), pp. 16–19.

CHAPTER NINE

Special Operations and Low-Intensity Conflict in 2010

Universally, the profession of arms is one of the most tradition-bound of all. Although military and naval officers are quick to grasp new technology, they invariably plan to use it in familiar ways. A new bomber, fighter, tank, cruiser, missile, submarine, or artillery piece is usually meant to improve on the model it replaces. But the fundamental task for these modern marvels is little different than that for their predecessors of forty to fifty years ago. There are many new weapons, few new roles. At the root of this is the fact that many, if not most, of the world's military organizations are planning to perform the identical mission they performed decades ago. Training of soldiers, sailors, and airmen has not fundamentally changed for some time, and neither has strategy. Tactics, due to the increased ranges and lethality of weapons, are a bit different, but even there change comes slowly. The profession of arms is one of the few where veteran grandparents can easily understand the roles their uniformed grandchildren are expected to play.

One of the basic reasons military forces change so little is that, unlike many other professions, practice is thankfully not continuous. It is quite common for many of the world's armed forces to see no combat, no clash of arms, for periods of thirty to forty years at a stretch. To update professional expertise, a doctor or lawyer only has to look at what happened last week, while the military or naval officer is often looking back a generation or more. Without a constant stream of experience, there is a natural tendency in the profession of arms to study the last war in preparation for the next one. Most successful generals and admirals read history and encourage their young officers to do so as well. That focus on the hard facts of past experience is reinforced by the knowledge that failure in the military profession can result in irreparable damage. A lawyer or doctor can lose a case, a bit of money, or a patient. The soldier, airman, or sailor can lose his life, or his country. Military or naval services are therefore experience-oriented. They are apt to bet on the realities of history, not new theory or new ideas.

Of course, the danger in relying on experience or the last war is that no two experiences, and no two wars, are ever the same. If there is a certainty in the profession of arms, it is that the next war will be different. There is an old saying in military circles: preparing for the last war ensures losing the next one. So the profession is always in a dilemma, attempting to hold on to the old while striving to foretell the new. The end result is that change comes slowly for military forces, but change does come.

Along with changes, there are some constants. History (experience) tells us that, from time immemorial, nations seek security, influence, and wealth. History also tells us that some nations will resort to war to obtain what they seek. Others may heavily arm to deter war. And some nations may create military alliances with nations of similar interests. History gives no indication of an end to war, an end to military forces, or even a successful and inexpensive method of avoiding war. Experience also teaches that warfare changes, and that it is best to be the progenitor of change rather than its victim.

The late 1990s and early 21st Century may bring as much change to warfare as the Napoleonic era brought in its wake. Alliances are changing, methods of warfare are changing, and there are fundamental shifts in diplomacy, new political trends, and major innovations in military technology coming to fruition. For those in the world's armed forces and those who earn their keep in the field of international security affairs, it is not enough to trust in past experience; current trends and future opportunities have to be examined.

Many of these trends have their beginnings in 1975, the start of the 20th Century's last quarter. The early part of that year brought the conquest of Cambodia and the Republic of Vietnam by totalitarian regimes. India, the world's largest democracy, was placed under a repressive state of emergency. Communist dictatorships triumphed in Angola and Mozambique. It was a year that saw an increasing use of hand-held weapons against the developed world's implements of mechanized warfare. But the year 1975 also saw Greece, Portugal, and Spain emerge as popular republics. For the first time, Western Europe became a solid, democratic entity. Elsewhere, representative government made more gains, eighteen other nations deciding on an electoral process. It seemed as though there was a global rush toward two camps, one composed of the democracies and the other devoted to totalitarian rule, command economies, and Marxist political philosophy. However, the era of the bipolar world was actually coming to an end.

Both superpowers began losing their grip on client states, and a world of more evenly shared power began to emerge. The Soviet Union and the United States experienced stagnant economies in the late 1970s, but that was only one reason for change. Of more importance, peoples and nations long allied with Washington and Moscow began to pursue their own interests at the expense of bloc unity. The two superpowers began to negotiate their disputes with each other, jointly moving toward a policy of *uti possidetis,* a policy under which "each holds its own." But some countries, particularly those with relatively free economies, had already begun a steady march toward economic and technological strength. Power is thus becoming more diffused and the worldwide bipolar structure is rapidly fading.

A pentapolar world is emerging. It is a world of five major powers, the Soviet Union, China, Japan, a growing and consolidating consortium of nations in Western Europe, and the United States, the latter just beginning on a path of economic consolidation with its neighbors, Canada and Mexico. If history is any guide, these five major northern hemisphere powers will seek security, influence, and favorable trade. They are now and will continue to be in competition. Four of these five powers will have great destructive means at their disposal, and long-time antagonisms exist between some of the five. Political and military competition will no doubt persist in the northern hemisphere. There will also likely be competition among the five for influence in the southern hemisphere. How they fare in this competition is in large measure dependent on time-tested measures of power: economic strength, political unity, diplomatic skill, military might, and technological prowess.

The results of this competition among the five centers of power will also depend on the wisdom of the chosen strategies. History teaches that the best strategy is likely to be one that uses inherent strengths aimed at a rival's natural weaknesses. Although most of the competition among the five will probably be in economic and political arenas, experience tells us that there will be military competition as well. In fact, sound strategies often depend on orchestrated efforts involving political, diplomatic, economic, and military programs.

Trends point to an increasing incidence rate of low-intensity conflict, and that level of violence will surely figure heavily in 21st Century strategy. But this does not mean that special operations forces will be primarily featured in these low-level wars and campaigns. Low-intensity conflict—armed conflict for political purposes not involving combat between regularly armed forces—may easily be conducted without special operations forces. There is no compelling reason for the terrorist, counterterrorist, peacekeeper, peacemaker, insurgent, or counterinsurgent to employ special operations forces. These forces do have skills that may be useful in low-intensity conflict situations, but none of the subcategories of low-intensity conflict demands their employment. Twenty-first-Century special operations forces may be better used in mid- and high-intensity conflicts. Growth in low-intensity conflict events and a shift in the employment of special operations forces is in part due to technological trends.

Technology in 2010

The technology of the 1990s will improve the insurgent's chances for success. Shoulder-fired anti-tank and anti-aircraft weapons favor the insurgent. Effective, hand-held anti-aircraft systems exist today and, by the mid-1990s, newer models of even greater effectiveness will be available. The technology for a powerful, light anti-tank weapon is just now emerging, and this weapon promises to severely undercut the fortunes of counterinsurgent forces. Clandestine communications equipment—small transmitters with great range and low intercept probabilities—should also be fielded in the mid-1990s. Low-bulk, lightweight clothing, body armor, and improved rations are rapidly coming into the market. These latter innovations will measurably increase the survivability and sustainability of small guerrilla units. The insurgents of the 21st Century will be far more elusive, self-reliant, and potent than those of the 20th Century.

Technological evolution also favors the terrorist. In some measure, this is because of natural trends in the world's industries. Economies of scale have produced ever larger nuclear power plants, fewer but more vital telecommunications centers, greater concentrations of highly toxic chemicals in pesticide and other manufacturing facilities, larger capacity liquified natural gas transports, and bigger passenger airplanes—all potential targets for the terrorist seeking the spectacular and tragic event. Newer tools of assassination such as the laser rifle and the man-portable anti-aircraft weapon are now and will continue creating nightmares for the security officer and the counterterrorist.

In creating more opportunities for the terrorist and insurgent, technological trends encourage a higher incidence rate for terrorism and insurgency, and thus more work for the counterterrorist and counterinsurgent. There are material innovations to assist in these tasks, but area anti-personnel weapons, improved baggage- and passenger-screening devices, and DNA visas are palliatives, not cure-alls.

To a lesser extent, emerging technology, together with skill and innovative techniques, favors the peacekeeper. The Israeli Defense Forces have proved that a well-manned land and air barrier can provide warning and some delay of outside attack, albeit at some expense. The same cannot be said of technological aids for the peacemaker. Future forces that have the onerous duty of bringing order out of armed chaos in some distant capital will likely have to rely on the traditional instruments, infantrymen acting in a police role. However, taken all together, technological trends underwrite growth in low-intensity conflict, not abatement.

Special Operations Forces in Low-Intensity Conflict

Some of the same technological factors that support a probable higher occurrence of low-intensity conflict in the 21st Century provide reasons to expect a greater utility for special operations forces, but not necessarily for the *same* reasons, and not necessarily in the field of low-intensity conflict. In point of fact, it can be expected that there will be an uneven, but unmistakable, decline in the worldwide employment of special operations forces in low-intensity conflict situations.

It is unlikely that governments will use their special operations forces for state-sponsored terrorist operations, and if the trend away from terrorist-initiated hostage barricade tactics continues, there will be a decline in terrorist-related special operations rescues, heretofore a primary role for several of these units. But there is merit in considering these forces for use in raids on terrorist training facilities, headquarters, and the offices of any terrorist-sponsoring element of a renegade government. Reprisal actions such as air raids and missile strikes bring brief satisfaction for an offended nation, but questions unfailingly arise. The government suspected of supporting terrorists will usually claim the air raid resulted in outrageous atrocities against children, schools, and hospitals, regardless of the facts of the case. What an air raid or missile strike cannot do is present proof. A well-led commando action, however, can.

Insurgency and counterinsurgency are largely Third World activities and, because of that, special operations forces of the major powers are not likely to see much direct employment in such conflicts during the 21st Century. There is a notable exception: Soviet Spetsnaz forces will probably be increasingly involved in counterinsurgency campaigns within Moscow's client states, and possibly nearer to home, within the western and southern rimlands of the Soviet Eurasian empire. But the days of U.S. Special Forces–led ethnic minorities in the Third World are probably gone forever. That era began to fade with the 1969 Guam Doctrine, relegating the American armed forces to assistance and advisory roles in Third World counterinsurgency campaigns. It probably died with the 1973 enactment of the War Powers Act that, in effect, altogether prohibited the use of low-visibility U.S. military operations in which armed American soldiers and airmen would directly participate in some other nation's war. The gradual withdrawal of Western forces from Third World conflicts has also seen fewer southern hemisphere actions by the British SAS. London's former colonies are increasingly independent of it in military affairs. However, there may well be a growth area for Western special operations forces in the field of Third World insurgencies—but in an indirect and supporting role.

Both technological and political trends point toward growing Western assistance to insurgent organizations in the Third World. Units such as the U.S. Special Forces and the British SAS are logical choices for training Western-favored Third World insurgent leadership cadres. There is likely to be a continuing and possibly expanding desire in the West to overthrow repressive southern hemisphere dictatorships, but it is unlikely that Western states will commit their own soldiers and airmen to these battlefields. The alternative—support through advice, training, and material aid—is a natural and attractive choice for Western political leaders and one that can be accomplished outside the conflict area. Cruise missiles, capable of being used in an insurgent resupply role, allow effective and efficient material support of guerrilla units without endangering the Western sponsors. Western technology and war stocks for its special operations forces, anti-aircraft and anti-tank missiles, communications equipment, rations and clothing, and medical supplies can easily enhance the combat effectiveness of an insurgent force. A supply system using cruise missiles is not only safe, but eliminates middlemen, often the cause of corruption and attendant political controversy within Western nations.

Emerging technology also points to an enhanced effectiveness in Western military psychological operations, operations that may be used in support of Third World insurgent movements. Smuggled high-impact videotapes, audiocassettes, and cruise missiles used to insert subliminal video messages into a target government's television broadcasts are some of the means that Western nations can employ in backing a guerrilla force and urban underground. All told, the West should have a capacity to heavily influence the outcome of a 21st Century Third World insurgency without direct exposure of any of its soldiers, airmen or intelligence operatives.

Counterinsurgency is a different matter. It is not a military endeavor that lends itself to effective assistance or support by "remote control." But Western-supported counterinsurgency campaigns are decidedly on the decline, and there is no identifiable trend or reason to expect growth here, at least for the West. While Western special operations forces were previously heavily committed to counterinsurgency campaigns, the legal prohibitions and policy changes of the 1970s make it unlikely that we would see a repeat performance, even were Western governments so inclined. The logical choice for 21st Century Western-sponsored counterinsurgency assistance efforts in the Third World is a commercial approach. Assistance and support of a Third World nation engaged in a counterguerrilla campaign could be far more efficiently achieved, far more politically acceptable, and far less painful if they were accomplished on a contractual basis. After all, these services involve advice and training, not combat. Leaving soldiers, sailors, and airmen at home and employing the services of a military assistance corporation composed of counterinsurgency experts would transfer a thankless task into the disciplines of the marketplace. If the corporation fails, or if its methods are judged to be inappropriate, its contract need only be canceled. Western flags, prestige, and honor would not be at stake. For Western military forces, counterinsurgency assistance should be thrown onto the rubbish heap of history. It was not a field of excellence for these otherwise fine organizations.

Moscow, however, has little experience in or inclination toward the use of a commercial approach. Spetsnaz has a future in the field of counterinsurgency. Third World Marxist states are increasingly besieged with guerrillas, and there is every indication that the phenomenon will not be restricted to southern hemisphere communist regimes. The Soviet Union is well experienced in counterinsurgency and, despite some setbacks in the 1980s, is by no means a defeated nation. The U.S.S.R. is likely to use its special operations forces often and well into the 21st Century.

There is also reason to expect that there will be distinct employment differences between Soviet and Western special operations forces in the field of stability operations. None of the world's special operations forces has any particular expertise or talents for peacekeeping missions, and no known trend or identifiable policy would suggest a future role of any significance for such units in that field. But there is precedence and logic in an expectation of expanded 21st Century employment of special operations forces in quickly executed peacemaking tasks.

Along with their growing counterinsurgency role, Spetsnaz leaders can look forward to an increasing number of episodes in which they are expected to reassert Moscow's control in conditions of urban unrest. This role is one that Spetsnaz is well suited for, since the force is basically a commando-trained light infantry organization. Opportunities for peacemaking employment are likely to be located within the Soviet Union proper, as well as within the borders of Moscow's western allies. Growing Soviet irredentism and unrest within the captive nations of the Soviet empire should make the future of Spetsnaz a lively one.

Spetsnaz has rarely been employed by itself in a stability operation, and the same is true for U.S. special operations forces. However, it is possible that the 21st Century could see somewhat of a change for the Americans, specifically the U.S. Rangers. The U.S. Army is fielding an entire Ranger Regiment, a more powerful force than was available during the Grenada operation in 1983. The Rangers were universally acclaimed for their performance in that small action, and for good reason. They are probably the finest light infantry force in the world. Twenty-first Century U.S. decision makers will have the option of turning over an entire stability operation to the Ranger Regiment, an action that would eliminate the awkward and confusing mix of units and headquarters that plagued the American performance in Grenada. While U.S. military airlift is not likely to grow appreciably during the 1990s, the air transport fleet that is programmed to be available, consisting of air-refuelable C-141s, will be perfectly capable of transporting and sustaining the entire Ranger Regiment, plus a contingent of supporting helicopters, to practically any point on the globe. The rapid employment of 1,700 highly motivated, skilled American Rangers constitutes a potent

U.S. capability to seize control and bring about order in some of the most unruly of situations.

Overall, there will probably be a slight decline in the use of special operations forces in low-intensity conflict. There may be counterterrorist raids by special operations forces, to gather proof of support for terrorists, and an increase in the use of Western special operations forces in the role of assisting Third World insurgent organizations. Spetsnaz will undoubtedly experience a plethora of counterinsurgency missions, just as American Rangers may find themselves in a number of stability operations. But these activities are not likely to rival the extensive low-intensity conflict involvement of special operations forces in the 20th Century. There is good reason to expect an overall decrease in the use of Western special operations forces in an counterinsurgency role during the early years of the 21st Century, and no cause to look for a significant contribution of these organizations in combating terrorism. Special operations forces have little to offer in peacekeeping missions. The expanding roles for special operations forces are not in low-intensity conflict. They are to be found on the higher end of the conflict spectrum.

Special Operations Forces in Mid-Intensity Conflict

The era of mechanized warfare may be coming to an end, and that phenomenon will have a major impact on the future of special operations forces. These units, among all the world's many categories of forces, are probably the least dependent on the mechanized implements of war. Other combat organizations either employ or plan to be supported by tanks, fighter aircraft, bombers, artillery, trucks, interceptor aircraft, extensive air defense systems, and mechanized infantry carriers. All these systems and vehicles are experiencing rapid escalation in procurement costs, as well as dramatic rises in operating and support expenses. Mechanized warfare was created to regain mobility during World War I. Yet mechanized warfare of the late 20th Century has increasingly resulted in stalemates. Not only are mechanized forces ineffective against insurgents, they do not seem to offer the policy maker the means to secure a political goal within the bounds of reasonable costs. Mechanized warfare is pricing itself out of the market.

Mechanized forces are primarily bought and maintained to execute mid-intensity conflict, combat between regularly organized military forces. Despite their recent record of producing tactical stalemates, they are considered to be offensive in character, offering the policy maker the advertised option of conquering an adversary. As such, their existence engenders a reaction in potential target nations—the creation and maintenance of yet more mechanized forces to counter the threat. The end result has been, is now, and will continue to be a costly arms race.

Special operations forces can exploit inherent vulnerabilities of mechanized forces, and perhaps even assist in their eventual elimination. These possibilities depend on the continuation and further development of the special operations leadership selection process, intelligent employment of such forces, and progress in weapons technology. Of the three, the leadership selection factor may be the most far-reaching.

Operating in the rear of enemy forces in small bands and without much supervision requires a high order of leadership. The characteristics of a successful leader under these conditions include determination, intelligence, self-reliance, and superb physical condition. The British SAS, the American Delta Force, and the U.S. Special Forces are all implementing a leadership selection process that is producing this type of officer, an expensive system that does not assume college graduates have the inherent attributes required for command on the battlefield. This system of leader selection has served well for the SAS and promises to do the same for its American counterparts, who adopted it in the 1980s. Unfortunately, the more senior U.S. special operations officers have not been selected by this process, yet they are the leaders who will guide the destiny of American special operations forces in the 1990s. If this system is recognized for what it probably is, a revolution in leader development, it will have an application beyond the confines of special operations forces. For the short term, it is needed to produce the leaders required for effective operations in the rear of an opponent's mechanized forces.

Mechanized forces depend on the existence of lines, because defensive lines must protect all of the impedimenta needed to support and sustain heavy formations. The rear of these forces is therefore vulnerable, and special operations elements are ideally organized, equipped, and trained to exploit the vulnerability. Exploitation can be realized—if special operations units are employed in a surveillance and selective interdiction role. The availability of cruise missiles, improved infiltration means, lightweight and low-bulk rations, better clothing, clandestine communications equipment, and target designators will greatly enhance the combat value of special operations forces operating in the rear areas of an opponent's army. Effective, shoulder-fired anti-aircraft missiles add to this value, and should bring the overall worth of special operations forces to high levels in the 1990s. In the late 1990s and early 21st Century, the value that special operations forces contribute on the battlefield could be even higher.

It seems probable that the late 1990s will yield a light anti-tank weapon that will defeat the modern tank, and that innovation, along with others, can provide the conditions for a major change in warfare. A territorial defense force with a special operations cadre could deter a potential aggressor and defeat him if he attacks, so long as the defenders have man-portable weapons that can destroy both tanks and aircraft, and as long as these forces are deployed in great depth throughout the area of conflict. This type of defense force is best organized using local militia elements supported from hidden caches of arms, communications equipment, and mines. It is best led by full-time professional soldiers who plan for and then train their charges during peace, and who are themselves trained to operate independently in the rear of an enemy's mechanized force. The cost of this type of defense is only a fraction of the costs involved in creating and maintaining mechanized forces. However, the greatest benefit of such a structure is that it has no offensive capability and thus cannot be rationally regarded as a threat by a neighboring state. The territorial defense scheme holds the promise of greatly reducing the worldwide arms race.

Special operations forces offer a number of attractive options to the mid-intensity conflict strategist, particularly the defense planners of the West. Surveillance and selective interdiction depend on a degree of technological prowess not yet attainable in Marxist states. Territorial defense is based on the unfearing ability of a government to widely disperse arms among its citizenry. Both are techniques founded on inherent Western strengths, and are also aimed at vulnerabilities that are common to the heavily mechanized forces of the Eastern bloc. Furthermore, if the procurement and operational costs of mechanized forces continue their uncontrolled upward spiral, both techniques may become attractive simply because of their affordability.

Special Operations Forces in High-Intensity Conflict

Another possible growth area for Western special operations forces is at the higher end of the conflict spectrum, high-intensity conflict. This area too is one where there are marked differences between Spetsnaz and its Western counterpart units. While the West has the potential of stimulating and supporting armed revolt within the Soviet empire, Moscow has no such hopes in the West. This Western capability, however, is a dangerous one and must be handled with great care. A Western force that is prepared to support resistance within the Soviet Union and the Warsaw Pact nations should exist, but there should be some credible means to demonstrate that the force will not be used in a period of peace. It should be for war use only, because its real worth lies in its deterrence value.

High-intensity conflict—combat involving the use of nuclear weapons—has no rational political purpose even if both opponents employ relatively small numbers of weapons. There is nothing to win, and the best outcome would yield only a diminished state of survival. Organizations, weapons systems, and forces that are expected to be employed in this form of warfare should therefore be configured so as best to contribute to deterrence. It does no good to conceal a high-intensity conflict capability. For that reason, a strong Western insurgency support force should be relatively open to Warsaw Pact inspection. It should provide Moscow yet one more reason to opt for peace.

Unfortunately, only the United States possesses a ready insurgency support force within N.A.T.O., and even that capability is in need of much expansion and improvement. With strong prospects for significant nuclear arms reduction agreements during the 1990s, the West will be in need of deterrent forces to complement its conventional forces, units likely to remain numerically inferior to those of the Warsaw Pact.

Low-Intensity Conflict in 2010

The obvious failure of Marxism, and the emergence of China, Japan, and a West European confederation as economic and political powers of great strength, have contributed to the end of the bipolar world and brought about a great reduction of tension between the two 20th Century superpowers. That, however, does little to reduce the incidence of war. The Third World is now, and probably will continue to be, the source of most of the world's armed conflicts. The major powers may occasionally intervene in these struggles, but the causes of the vast majority of wars will undoubtedly continue to be found in the world's poorer nations. And the overwhelming majority of those wars are likely to be low-intensity conflicts.

Another characteristic of the year 2010, familiar to those who will have lived through the last quarter of the 20th Century, is that most of the world's low-intensity conflicts will probably be insurgencies. Terrorism, in and of itself, is a weak reed when it comes to effecting political change. On the other hand, governments have been brought down by insurgents. Insurgency is the strongest form of low-intensity conflict and, as mentioned above, the insurgent is likely to be an even more potent threat to governments in the late 1990s and early years of the 21st Century.

One aspect of insurgency that promises to be a bit different in the year 2010 has to do with a shift in demography. The continued movement of Third World populations to cities makes it probable that urban underground organizations will constitute a growing percentage of insurgent movements, a change taking place at the expense of rural guerrilla components. This will place a greater burden on police activities for the 21st Century counterinsurgent and bring about a slight decrease in the role of military counterguerrilla forces operating in the countryside.

Some insurgencies may not involve major power involvement, but some special conditions will likely compel outside intervention in others. These conditions include threats to northern hemisphere access to Third World raw materials, particularly oil. Additionally, there is an age-old staple of conflict causation, the rivalry between major powers for influence among lesser states. Both Moscow's Third World clients and those nations that mimic the Soviet style of government are likely to request military assistance from the U.S.S.R. in conducting counterinsurgency campaigns. And it is precisely this phenomenon and not ideology per se that may stimulate Western aid to an insurgent organization.

North-South relations in 2010 should also be marked by another familiar 20th Century pattern, Third World-spawned terrorist attacks against Westerners. Whether sponsored by Third World governments or by radical Third World political groups, these troublesome and tiresome events will certainly continue, although there should be a general change in their character. The likely tool of the 21st Century terrorist will be the well-placed bomb. However, there is also a likelihood that the terrorist may no longer be satisfied with explosives alone. Chemical and possibly biological dispersants can be expected in the coming years. Although there will certainly be a continuation of the North-South terrorist and counterterrorist battle, Third World citizens will probably continue to constitute the majority of victims.

However, 21st Century North-South relations may differ from 20th Century norms. The major powers may be more willing to use force in the Third World. In part, that change arises from the decline in bipolar ideological competition. U.N. Security Council actions, long hampered by distrust between the Soviet Union and the United States, are not likely to be stifled as they previously were. While rivalry between the major powers will exist, there should be a growing number of situations wherein the developed world takes concerted action to achieve stability in some corner of the Third World. For that reason, the world can take some heart in facing the 21st Century—there will probably be more peacekeeping operations.

Western Decisions for 2010

Even tomorrow holds uncertainties. A major, unexpected event can provoke a novel reaction and forever change the course of nations. An individual leader can chart wholly new policies sparking a series of counteractions that, when taken all together, have global political, economic, and military impact. A disciplined, apparently predictable technological program can suddenly yield a surprise outcome that results in new departures in human life-styles, the way man acquires knowledge, or man's subsequent actions.

But "great events" are rare. Even if they do occur, there is always the possibility that people and nations will continue to go about their activities in a characteristic way, ignoring what should or might happen, doing what they intended in the first place. To passively await the big event is surely to surrender one's fate to others, because people and nations are constantly planning, projecting, and in effect attempting to determine their own destinies. In an ever more interdependent world, a nation's condition at a given point in time is more likely to be influenced by rational plans than by unexpected events. And those plans can be the plans of a hostile nation. Given a choice, it is better to plan one's own destiny, and that is the business of government. Officials are paid to plan; they are paid to decide.

Western officials with the duty of safeguarding the fate of the West have a number of opportunities to influence the course of future events. Whatever their actions may be, the policies and programs of the 1990s will have an effect on the course of the West in the year 2010. Planning for the future is not risk free, but neither is an attempt to maintain the *status quo*. While it is not wholly foolish to take counsel of fear and be reluctant to change, it will serve the West better if its officials look to the opportunities. Opportunities are exploited by taking initiatives, and successful initiatives are the tools to determine one's own fate. Some of the available initiatives are in the separate fields of low-intensity conflict and special operations.

A rise in peacekeeping operations in the late 1990s and early 21st Century appears to be likely, but the community of nations is not yet prepared to handle these growing tasks. A continuation of the world's current peacekeeping procedures will result in yet more deployments of poorly trained, ill-motivated military forces, organizations that regard peacekeeping as a secondary and unwelcome duty. It would clearly be helpful to bring the Military Staff Committee of the U.N. Security Council to life. Japan, an economic giant and the single major power with relatively minimal security expenditures, should be encouraged to take over some of the world's peacekeeping burdens. Japan could fund professionally trained commercial organizations that are wholly devoted to peacekeeping, organizations whose performance would be subject to the disciplines of the marketplace. Japan could use its considerable managerial talents to make peacekeeping effective. Japan could also make investments in training, technology, and investigative techniques to help make peacekeeping a more thoroughly competent profession. Peacekeeping is simply too important to be left to amateurs, casual planning, and hastily assembled *ad hoc* solutions.

There is no indication that terrorism will somehow fade away in the future, but there are opportunities for the community of nations to undercut the strategy of most terrorist organizations. A flat prohibition on the publication or broadcast of terrorist demands, terrorist pronouncements, or even the names of terrorist organizations would deprive many of these groups of their *raison d'être*. Obviously, this initiative will be difficult to realize in some democratic states, where the power of the media intimidates politicians, but the potential rewards are high. There also needs to be a concerted international effort to place tracer codes in explosives, compile and maintain a worldwide data base on weapons transfers, and tighten identification procedures for international travel. Technology offers great opportunities in screening passenger baggage and individuals so initiatives in this field must be funded and supported.

Western security officials and military leaders of 2010 will probably be involved in a number of stability operations, but there are few current innovations in this field, leading one to expect a lack of improvement in future performance. West European leaders will likely have need of substantial military airlift in future stability operations. Thus an immediate European investment in capable modern military transport aircraft would not come too soon. As for American leaders, there seems to be merit in adding peacemaking operations to the Rangers' menu of missions and initiating plans to give the Ranger regimental commander unfettered control of some of these operations, avoiding the confused and awkward command arrangements of the past.

The West will also face some counterinsurgency contingencies in the future, although their number is likely to be small. The best course for the West is to cut its military losses; delegate military training, advisory, and support duties in the arena of counterinsurgency to a commercial endeavor; and support one or more multinational corporations composed of the few experts available. Counterinsurgency is not an enterprise in which the West should risk the prestige of its armed forces or Western political honor. In the end, the outcome of a counterinsurgency campaign is almost always contingent on the performance of a Third World government, not on Western assistance or Western actions. These campaigns are occasionally necessary endeavors, but they are efforts that can be put at some distance from Western prestige and should be. Counterinsurgency is not now, nor is it likely to be in the future, a field in which the West excels.

Insurgency should be viewed in a different light. For the West, this field is growing, not declining, and although it is not likely that Western military organizations will often be directly involved in overthrowing Third World autocratic regimes, it is quite possible the West will desire to support and assist some future insurgent movements. The expertise, war stocks, and training abilities of Western special operations forces can make an important contribution in these campaigns. Thus the West should enhance its special operations forces so as to better its chances of influencing the outcomes of 21st Century low-intensity conflicts in the Third World.

Special operations forces will also be useful in mid-intensity conflict, more so than ever before. The vulnerability of satellites makes it imperative to provide a backup capability to gather information in the enemy rear. Properly equipped, special operations forces can provide that and more. These forces can direct rear-area missile strikes, severely crippling an opponent. Western decision makers should closely monitor and support programs to give special operations forces cruise missile resupply systems, meteor-burst communications systems, laser designators, adequate lightweight and low-bulk clothing, body armor, and lightweight rations.

It is even more important to examine the possibilities of a territorial defense system. This type of defense has the potential of defeating a mechanized attacker. Special operations forces are ideally suited for providing the professional cadre for a territorial militia. The territorial defense concept holds the promise of drastically curtailing the arms race, thus providing some democracies with an adequate defense at reasonable costs.

However, the greatest contribution that special operations forces can make to the West probably lies in the field of high-intensity conflict and specifically in the realm of war deterrence. Western decision makers have an opportunity at hand to turn Soviet irredentism and East European dissidence into yet another reason for Moscow never to consider an armed conflict with the West. To realize that capability, the West must considerably strengthen special operations forces dedicated to European contingencies, heavily invest in language training, and fund a development program to bring about a low-observable transport aircraft that has a good chance of penetrating Warsaw Pact air space. In order to produce the full deterrent value of such a force, Soviet authorities must be made fully aware of its capabilities. This type of capability should be under the command of Western military leaders and, unlike Spetsnaz, have no control links to any national intelligence service.

LIC 2010: A Western Way of War

Four hundred years before the birth of Christ, Plato remarked, "Only the dead have seen an end to war." Perhaps he was right—war may be inevitable. But since mankind makes wars, it is reasonable to believe mankind has the power to stop them, or at least control their horrible results. That, of course, has been tried before. During the dark days of the last global conflict, the Security Council of the emerging United Nations was designed as a body to maintain the peace then being won with so much blood. The council was to be composed of representatives from the major powers. There were five of them. It was to be a pentapolar world. Although the concept failed in the 20th Century, it might work in the 21st. After all, while man is heavily influenced by continuity, man is also a creature of change.

Plato would be very interested in our world, but not necessarily in the ingenious ways we have devised to inflict violence on each other. His great field of expertise lay in ideas of governance. But the fact that modern democracies do not fight one another would not escape his keen powers of observation. And, it is true, wars are a phenomenon of autocratic states, totalitarian regimes, peoples that have the misfortune to suffer a dictator, or those who fear the yield of an honest ballot box. Unfortunately, democracies must not only survive in a world with these other nations and peoples, but occasionally risk all in armed conflict with them.

So, democratic nations must create ways to fight and win. The best strategies use inherent strengths to attack an adversary's vulnerabilities. A territorial defense system and an offensive capability to stimulate and support insurgency in an enemy's homeland both play to democratic strengths and are directly aimed at the natural vulnerabilities found in the likely adversaries of the West. In essence, both concepts are based on the will of the majority. They are less expensive than mechanized warfare. They provide offensive and defensive means to survive and win. Neither concept involves mass destruction weapons. Territorial defense and a ready structure to foster insurgency both depend on strong and capable special operations forces. Both concepts are based on the idea of achieving political goals through techniques now considered to be in the realm of low-intensity conflict.

Today, special operations and low-intensity conflict (LIC) are properly considered as two separate subjects, but that may change. In the 21st Century, there could be a melding of the two. That merger could make the 20th Century distinction in levels of conflict obsolete. For the democracies, there may be only one level of conflict. In 2010, low-intensity conflict could become the Western way of war.

Selected Bibliography

Although there are many ways to approach the future of special operations and low-intensity conflict, a researcher might begin with some finding aids and then delve into the nature of war itself, concentrating on how wars are begun. That could be followed by an examination of existing special operations units and current thinking on low-intensity conflict. Since the latter is normally associated with the poorer nations, a look into Third World studies would be of assistance. Armed with such background material, a plunge into the world of future studies is recommended, beginning with some sound monographs on tomorrow's international security environment. Finally, there are some interesting books and articles dealing with the future of warfare that warrant consideration.

Regardless of the approach, a good starting point is provided by three bibliographies, all products of the fine libraries at Carlisle Barracks, Pennsylvania. See U.S. Army War College Library, *Low Intensity Conflict: A Selected Bibliography* (Carlisle Barracks, PA, 1986) and *Special Operations: A Selected Bibliography* (Carlisle Barracks, PA, 1989). A more extensive listing of sources, largely historical, is available in Professor Claude C. Sturgill's Special Bibliography Number 26, *United States Army Special Operations and Low Intensity Conflict, 1940–1986* (Carlisle Barracks, PA: U.S. Army Military History Institute, 1989).

The Origins of War

Perhaps a bit dated, but nonetheless a staple for an examination of the nature of wars, Quincy Wright's second edition of *A Study of War* (Chicago: University of Chicago Press, 1965) provides a fundamental understanding of armed conflict at all levels of intensity. A.J.P. Taylor's *How Wars Begin* (New York: Atheneum, 1979) is a short, somewhat offhand view of conflict causation. It should be balanced with a more disciplined treatment: John G. Stoessinger, *Why Nations Go to War* (New York: St. Martin's Press, 1985). A good book that describes the evolution of war in the context of technology is Robert L. O'Connell's *Of Arms and Men: A History of War, Weapons, and Aggression* (New York: Oxford University Press, 1989). An analytical treatment of wars is found in G.D. Kaye et al., *Major Armed Conflicts: A Compendium of Interstate and Intrastate Conflict, 1720–1985* (Ottawa: Department of National Defense, 1985).

Special Operations Forces

An examination of all the world's special operations forces does not exist and should probably be undertaken. The approach in these pages is to study the three national models that most such units are patterned after: British, American, and Soviet. For the British model, see Tony Geraghty, *Inside the SAS* (New York: Ballantine Books, 1982), perhaps the only description of the unit done with the assistance of the regiment. Both Soviet and American special operations forces are detailed in an excellent study done by Colonel John M. Collins of the U.S. Congressional Research Service, *Green Berets, Seals and Spetsnaz: U.S. and Soviet Special Military Operations* (Washington, D.C.: Pergamon-Brassey's, 1987). Spetsnaz is also described in Viktor Suvorov's *Spetsnaz: The Inside Story of Soviet Special Operations Forces* (New York: W.W. Norton Co., 1987). However, it would be advisable not to put too much faith in the mysterious Suvorov (whoever he is), so his offering should be balanced by consulting David Isby's works on the subject, available in several periodicals and studies.

Low-Intensity Conflict

Brigadier Kitson's slim volume is in need of updating, but it is still recommended, if for no other reason than to follow the logic and rationale for various categories of conflict. See Frank Kitson, *Low Intensity Operations: Subversion, Insurgency, Peacekeeping* (Harrisburg, PA: Stackpole Books, 1971). It is a good starting point for the novice. In this field, there are a number of thoughtful studies produced by joint working groups of the U.S. Army and Air Force. See U.S. Army and U.S. Air Force Low Intensity Conflict Study Group, *Joint Low Intensity Conflict Project Final Report* (Ft. Monroe, VA: U.S. Army Training and Doctrine Command, 1986). See also the studies on low-intensity conflict produced by the joint services center for the study of low-intensity conflict. Almost every facet of this end of the conflict spectrum is addressed by these studies. They are too numerous to mention here, but one of the better ones, Major William H. Thornton's "Modern Terrorism: The Potential for Increased Lethality" (Langley Air Force Base, VA: Army–Air Force Center for Low Intensity Conflict, 1987) is particularly useful.

A fundamental text on terrorism has been written by Francis M. Watson, *Political Terrorism: The Threat and the Response* (New York: Robert B. Luce Co., 1976). Terrorism is well detailed in an anthology that contains a large documentation section: Uri Ra'anan et al., *Hydra of Carnage: International Linkages of Terrorism, The Witnesses Speak* (Boston: D.C. Heath, 1986). An expert's view is found in Brian Jenkins's *International Terrorism: The Other World* (Santa Monica, CA: The Rand Corporation, 1985). An approach to counterterrorism is provided by Noemi Gar-or, *International Cooperation to Suppress Terrorism* (New York: St. Martin's Press, 1985).

Peacekeeping from a United Nations standpoint is presented in Major General Indar Jit Rikhye's *The Theory and Practice of Peacekeeping* (New York: St. Martin's Press, 1984). A previous study is also useful. See Inis L. Claude's *Swords into Plowshares* (New York: Random House, 1972). Peacemaking operations, at least the air aspects, are addressed by Major Bradley L. Butler. See his "Planning Considerations for the Combat Employment of Air Power in Peacetime Contingency Operations" (Langley Air Force Base, VA: Army–Air Force Center for Low Intensity Conflict, 1988).

Modern insurgency and counterinsurgency are described in Ian F.W. Beckett and John Pimlott, eds., *Armed Forces and Modern Counterinsurgency* (New York: St. Martin's Press, 1985) and in Richard H. Shultz, Jr., *The Soviet Union and Revolutionary Warfare: Principles, Practices, and Regional Comparisons* (Stanford, CA: Hoover Institution Press, 1988). Professor Shultz has also edited another work concerning insurgency and counterinsurgency: *Guerrilla Warfare and Counterinsurgency: U.S.-Soviet Policy in the Third World* (Lexington, MA: Lexington Books, 1989).

Third World Studies

There are any number of good works on the Third World, but several are notable for concentration on military aspects. Ruth Leger Sivard's *World Military and Social Expenditures, 1987–1988,* 12th ed. (New York: World Priorities, 1987) focuses on Third World armaments. It should be handled with some care, since the text has a political message (guns make people bad); there is, however, no reason to believe the statistics are flawed. Another view is contained in A.F. Mullins, *Born Arming: Development and Military Power in New States* (Stanford, CA: Stanford University Press, 1987). A book on Third World economics that has rightfully captured much interest, Hernando de Soto's *The Other Path: The Invisible Revolution in the Third World* (New York: Harper and Row, 1989), should be studied, even though its data are derived from only one Latin American nation. For a dismal, but honest, view of the difficulties in aiding the poorer nations, see U.S. Agency for International Development, "Development and the National Interest into the 21st Century" (Washington, D.C.: USGPO, 1989).

The Future International Security Environment

Possibly the best single monograph on the projected security environment for the world in the year 2010 is Charles W. Taylor's *A World 2010: A Decline of Superpower Influence* (Carlisle Barracks, PA: Strategic Studies Institute, 1986). A number of good studies were produced in 1988 by the Commission on Long Term Strategy headed by Fred C. Iklé. The main report is titled *Discriminate Deterrence* (Washington, D.C.: USGPO, 1988). David M. Abshire's *Preventing World War III: A Realistic Grand Strategy* (New York: Harper and Row, 1988) is recommended. Also, see Perry M. Smith et al., *Creating Strategic Vision: Long Range Planning for National Security* (Washington, D.C.: National Defense University Press, 1987). Among the Iklé Commission's subreports, Andrew W. Marshall and Charles Wolf's "Sources of Change in the Future Security Environment" (Washington, D.C.: Department of Defense, 1988) is very useful.

Future Warfare

For tomorrow's land warfare, see Chris Bellamy, *The Future of Land Warfare* (New York: St. Martin's Press, 1987). An optimistic view of technological progress in the 21st Century is contained in Steven M. Shaker and Alan R. Wise, *War Without Men: Robots on the Future Battlefield* (McLean, VA: Pergamon-Brassey's, 1988). For a view of air warfare in the future, see Major General Perry M. Smith's "Air Battle 2000 in the N.A.T.O. Alliance: Exploiting Technological Advances," *Air Power Journal* (Winter 1988). Finally, a thought-provoking look at the future is found in Richard E. Simpkin's *Race to the Swift: Thoughts on Twenty-First Century Warfare* (London: Brassey's Defence Publishers, 1985).

Index

About the Author

Colonel Rod Paschall, U.S. Army (Ret.), is a 1959 graduate of West Point and holder of an M.S. degree from George Washington University and an M.A. from Duke. During his army career, he served in eleven foreign countries, mostly as a Special Forces officer. He saw numerous combat assignments, served two tours in research and development positions, was a plans officer in the special operations division of the Office of the Joint Chiefs of Staff, and commanded the U.S. Army's Delta Force. Col. Paschall's last assignment in uniform was as the director of the U.S. Army Military History Institute. The author of *The Defeat of Imperial Germany, 1917–1918* (Algonquin), he lives in Carlisle, Pennsylvania, where he is working on a book about the Second Indochina War (forthcoming from Avery in 1991).